The Compromise of LIBERAL ENVIRONMENTALISM

Steven Bernstein

COLUMBIA UNIVERSITY PRESS • NEW YORK

COLUMBIA UNIVERSITY PRESS
Publishers Since 1893
New York   Chichester, West Sussex
Copyright © 2001 Columbia University Press
All rights reserved

Library of Congress Cataloging-in-Publication Data

Bernstein, Steven F.
  The compromise of liberal environmentalism / Steven F. Bernstein.
    p.   cm.
  Includes bibliographical references and index.
  ISBN 0-231-12036-2 (cloth : alk. paper) —ISBN 0-231-12037-0 (pbk. : alk. paper)
    1. Environmental policy. 2. Sustainable development. I. Title.

  GE170 .B47 2002
  363.7'05—dc21   2001028016

Casebound editions of Columbia University Press books are printed on
permanent and durable acid-free paper.

Printed in the United States of America

c 10 9 8 7 6 5 4 3 2 1
p 10 9 8 7 6 5 4 3 2 1

The Compromise of LIBERAL ENVIRONMENTALISM

CONTENTS

LIST OF ACRONYMS

AGGG	Advisory Group on Greenhouse Gases
BCSD	Business Council for Sustainable Development
CBD	Convention on Biological Diversity
CDM	Clean Development Mechanism of the Kyoto Protocol
CFCs	Chlorofluorocarbons
CHP	Common Heritage Principle
CIDA	Canadian International Development Agency
CITES	Convention on International Trade in Endangered Species of Wild Fauna and Flora
CTE	Committee on Trade and Environment of the WTO
EC	European Community
ECOSOC	United Nations Economic and Social Council
ENGO	Environmental Nongovernmental Organization
EPA	Environmental Protection Agency (United States)
EU	European Union
FAO	United Nations Food and Agricultural Organization
FCCC	Framework Convention on Climate Change
FRG	Federal Republic of Germany (West Germany)
FSC	Forest Stewardship Council
G-7	Group of Seven
G-77	Group of Seventy-Seven
GARP	Global Atmospheric Research Programme
GATT	General Agreement on Tariffs and Trade (WTO after 1994)
GCC	Global Climate Coalition
GDP	Gross Domestic Product
GDR	German Democratic Republic (East Germany)

GEF	Global Environment Facility
GEMS	Global Environmental Monitoring System of the U.N.–sponsored Earthwatch Network
GHG	Greenhouse Gas
IBP	International Biological Programme
ICC	International Chamber of Commerce
ICSU	International Council of Scientific Unions
IGBP	International Geosphere-Biosphere Programme
IHDP	International Human Dimensions Programme
IIASA	International Institute for Applied Systems Analysis
IMF	International Monetary Fund
IPCC	Intergovernmental Panel on Climate Change
IPE	International Political Economy
IUCN	Word Conservation Union—renamed 1990, formerly International Union for the Conservation of Nature and Natural Resources
JI	Joint Implementation—also knows as AIJ: Activities to be Implemented Jointly
LMOs	Living Modified Organisms
MAB	Man and the Biosphere Programme
NIEO	New International Economic Order
NGO	Nongovernmental Organization
ODA	Official Development Assistance
OECD	Organization for Economic Cooperation and Development
PPI	Progressive Policy Institute
PPMs	Process and Production Methods
PPP	Polluter Pays Principle
PRC	People's Republic of China
SCOPE	Scientific Committee on Problems of the Environment of ICSU
SSTs	Supersonic Transports
UNCED	United Nations Conference on Environment and Development, 1992
UNCHE	United Nations Conference on the Human Environment, 1972
UNCLOS III	1982 United Nations Convention on the Law of the Sea
UNCSD	United Nations Commission on Sustainable Development
UNCTAD	United Nations Conference on Trade and Development
UNDP	United Nations Development Programme
UNEP	United Nations Environment Programme
UNESCO	United Nations Educational, Scientific, and Cultural Organization
UNGA	United Nations General Assembly
UNGASS	United Nations General Assembly Special Session to Review Implementation of Agenda 21
WCED	World Commission on Environment and Development (Brundtland Commission)
WCS	World Conservation Strategy
WCRP	World Climate Research Programme
WMO	World Meteorological Organization
WTO	World Trade Organization
WWF	World Wide Fund for Nature renamed in 1990—formerly World Wildlife Fund

PREFACE AND ACKNOWLEDGMENTS

THE GERM OF THE IDEA that eventually evolved into this book came on the eve of the 1992 Earth Summit in Rio de Janeiro. It was a time of great hope for people who shared a concern over the fate of the global environment, but tempered by uneasiness over the course the summit set for understanding and addressing these problems. Almost 10 years later, the legacy of those ideas and decisions—a legacy I label the "compromise of liberal environmentalism"—remains powerfully in place. It can be found in the ideas that undergird many of the most important and far reaching international environmental treaties such as on climate change and biodiversity, in the policies and programs of international environment and development organizations ranging from the United Nations Environment Programme to the World Bank and throughout the United Nations system as a whole, and in the way powerful non-environmental organizations such as the World Trade Organization—which increasingly finds itself making decisions with potential environmental consequences—try to respond to the demands to accommodate environmental and development concerns in their decisions and policies. Even many nongovernmental organizations rest their proposals for action on this framework, as do many states in their domestic policies.

Given the potential impact of these institutions and policies for the fate of the planet's ecological systems and human life and health, I thought it crucial to uncover the reasons why states and international institutions treat global environmental problems the way they do, and the underlying forces of world politics that pushed global environmental governance in this direction. Only with this knowledge did it seem possible to make informed judgments about the prospects for future action, or provide a basis on which to critically assess and debate whether there is a need for deeper change. This task seems even more important today—given the disappointment many feel, despite some notable successes, over progress on the world's most severe environmental problems since 1992.

The second inspiration for this book came from a frustration, typical among many of my fellow graduate students at the time, with most of the literature on international institutions. It seemed to forget that international institutions are not simply a vehicle through which states cooperate, but that the cooperation they enable is for some purpose or goal. Purposes and goals and the politics that drives them were simply left out of most International Relations scholarship, as if all that mattered was whether an institution formed or not. What it actually did (or did not do) mattered not at all. A new literature in International Relations—which now falls under the label 'constructivism'—challenged the discipline by bringing politics and common or collective purposes back into the analysis of international institutions. However, it faced heavy criticism initially for being too concerned with big theoretical questions at the expense of adding to our knowledge of important and pressing problems in world politics. My hope is that this book is one in a growing list that responds to this criticism.

I owe a great many people thanks for support, encouragement, constructive criticism, and even inspiration as I developed my ideas. At University of Toronto, Janice Gross Stein, Ronald Deibert, Louis Pauly, Robert O. Matthews, and Craig Scott asked the tough questions and pushed and prodded my work to reach beyond its initial limitations. David Welch merits special mention for what in hindsight must have been a painstaking task; his willingness to read and comment in detail on multiple versions of each portion of what would later become a draft manuscript.

A great many other people helped along the way through thoughtful discussions, insights, and comments on the research that eventually coalesced into this book. They include Thomas Biersteker, Ben Cashore, Dorinda Dallmeyer and other participants in the 1996 ACUNS/ASIL Summer Workshop on "The Role of Governmental, Intergovernmental, and Nongovernmental Institutions in Global Governance" at Brown University, Andrew Deutz, Elizabeth DeSombre, Joanne Gestrin, Thomas

Homer-Dixon, Richard Herrmann, Andrew Hurrell, Rhona Leibel, Karen Litfin, Miriam Lowi, Richard Matthews, Don Munton, Shaun Narine, Norrin Ripsman, and Lisa Young.

I especially want to acknowledge my intellectual debt to James Busumtwi-Sam whose collaboration on a related project greatly helped my thinking on many theoretical aspects of this book. Our work together at times became so intertwined that he deserves some credit for whatever intellectual contribution is made here. Of course the blame for any errors of fact or judgment are solely my own.

Other people took time out of their busy schedules to discuss their experiences and issues about which they knew far more than me. While the formal interviews directly quoted are listed in my bibliography, I also thank Louise Comeau, Alden Meyer, Peter Timmerman, Jeffrey Watson, and Doug Whelpdale, among others, who talked to me about their research or participation in various international environmental negotiations. I am also grateful to Peter Berry who shared his interview notes on ozone negotiations and research with me. I especially want to thank Louise Comeau and Jennifer Morgan, of the Climate Action Network, for helping me arrange to attend the first Conference of the Parties for the Framework Convention on Climate Change in Berlin, and for giving me full access to CAN's strategy sessions and information.

An Eco-Research Doctoral Fellowship from the Tri-Council Secretariat of Canada provided the major source of funding for research. Generous financial support from the Halbert Exchange allowed me to spend a year at the International Relations Department at Hebrew University where my many discussions with Emanuel Adler on constructivism and epistemic communities challenged me to consider more carefully key elements of my arguments. Although we did not always agree, his intellectual and personal mentorship were without rival. Similarly Ned Lebow, director of the Mershon Center of Ohio State University equally provided intellectual enrichment and friendship during the final stages of writing the draft manuscript. Generous postdoctoral funding from the Mershon Center and the Social Sciences and Humanities Council of Canada provided me the time and ideal writing environment away from other university duties to see the project through to completion.

All this would have been for nought without the commitment and support for the project from Kate Wittenberg at Columbia University Press. The comments of the two anonymous reviewers she chose helped make the final manuscript a much stronger and more coherent piece of work. I am also grateful to Leslie Bialler and the rest of the staff at Columbia for their efficiency and professionalism in ushering a first-time author through the arcane editorial process.

Parts of the argument and evidence developed further here appeared first in "Ideas, Social Structure, and the Compromise of Liberal Environmentalism." European Journal of International Relations 6 (4) (December 2000):464–512, and are reproduced with permission from Sage Publications and the European Consortium for Political Research.

This book would never have been written without the unwavering support and inspiration of my partner and wife, Linda White. Her painstaking scrutiny of virtually every word I have written related to this project made it a far better product than I ever could have produced on my own. She constantly pushed me to clarify my ideas, asking the fundamental questions about my work that others would not. I dedicate this book to her.

Steven Bernstein
Toronto, Canada
January 2001

The Compromise of LIBERAL ENVIRONMENTALISM

Chapter 1
INTRODUCTION

We cannot say with certainty how much longer mankind [sic] can postpone initiating deliberate control of his growth before he will have lost the chance for control. We suspect on the basis of present knowledge of the physical constraints of the planet that the growth phase cannot continue for another one hundred years. Again, because of the delays in the system, if the global society waits until those constraints are unmistakably apparent, it will have waited too long.

—*The Limits to Growth* 1972

The concept of sustainable development does imply limits—not absolute limits but limitations imposed by the present state of technology and social organization on environmental resources and by the ability of the biosphere to absorb the effects of human activities. But technology and social organization can be both managed and improved to make way for a new era of economic growth.

—*Our Common Future* 1987

FEW TRULY GLOBAL concerns have held the potential to transform substantially the nature of global politics and society. Contenders might include the fear of nuclear annihilation or advances in technology and telecommunications. The former arguably has transformed the nature of conflict between the major powers, while the latter have made possible exponential increases in economic transactions across vast distances, enhanced the spread of culture, and enabled vast changes in the patterns of interaction between a wide range of actors on the global stage.

Looking back thirty years, one might have predicted that the concern over the state of the global environment could similarly transform global politics. Responses to such concerns have called for a whole new notion of planetary rather than national security and thrown into question the assumption of competing interests of states or the ability of such units, or the sovereign state system they comprise, to manage global problems. Furthermore, a growing awareness of environmental problems and ecological interdependencies has led many to question the wisdom of conducting global economic relations as if they were independent from the ecological systems that sustain life on the planet.

The early ideas that informed international attempts to manage the Earth's resources supported such transformations. The philosophical statement of planetary concern commissioned for the first global environmental conference—the United Nations Conference on the Human Environment in 1972—included calls for a "loyalty to the Earth" that recognized planetary interdependence of all life, the adoption of global (as opposed to national) responses to environmental problems, and massive changes in over-consumptive lifestyles of the wealthy. *Only One Earth*, as it was called, also criticized existing international institutions for lacking a sense of planetary community and commitment (Ward and Dubos 1972). High-profile studies such as *The Limits to Growth* took an even tougher stand against overconsumption and warned that growth in population and production could not continue on course without leading to the collapse of social and economic systems (Meadows et al. 1972). No one expected revolutionary changes to occur overnight, but an assumption continues to prevail that as the international community pays more attention to environmental problems, we will move gradually toward a more ecological understanding of our world and humankind's place in it. At the least, our responses to environmental problems themselves will lead us in an ecological direction.

This book examines whether indeed that is the case. It does so by detailing how international concern for the global environment moved from these initial formulations to the current concern with "sustainable development," and what form of international governance "sustainable development" entails. This evolution of environmental governance takes on added significance when one considers that environmental issues finally reached the mainstream of international relations in the early 1990s only when they took this form.

Whether or not sustainable development constitutes a truly transformative idea, international lawyers and political scientists note that the 1992 Earth Summit in Rio de Janeiro institutionalized ideas associated with this new conception of environmental governance. Some call it a "paradigm shift" to a new international law of sustainable development from previous formulations of both an international law of the environment and of development.[1] Others argue that the Earth Summit "succeeded in formulating an umbrella regime in the field of sustainable development" that will continue to shape specific responses to environmental problems well into the future (Sjöstedt et al. 1994:5). These institutionalized ideas arguably embody the most significant shift in environmental governance over the last thirty years. Not only did they bring environmentalism into the mainstream of international governance, but they did so by reformulating environmental concerns in the context of a liberal international economic order. In that way, sustainable development does mark the institutionalization of environmental concern, but not as originally envisaged.

Instead, the compatibility of environmental concern, economic growth, the basic tenets of a market economy, and a liberal international order is now conventional wisdom among many policy makers, diplomats, and a large number of nongovernmental organizations throughout the world. It is easy to forget that this formulation of the environmental problematique differs substantially from those dominant when the first concerted efforts at wide-scale global responses to environmental problems began in the late 1960s and early 1970s. From the perspective of those earlier efforts, focused on the negative environmental consequences of unregulated industrial development and suspicious of economic growth, the shift in environmental governance is a remarkable and a largely unforeseen departure. Why, then, when the international community finally took environmentalism seriously, was it only considered in the context of an economic program that not only encouraged growth, but actually demanded it? Why did international environmental governance evolve into what I will call the compromise of "liberal environmentalism?"

These questions are too often overlooked in academic and policy work overwhelmingly focused on the quest to design better institutions to manage the Earth's resources or respond to immediate and pressing problems.[2] This omission also points to a serious gap in the literature on international institutions more generally. Dominant strands of the rational institutionalist "regime" literature, for example, generally ignore the question of which values cooperative outcomes promote, because they focus primarily on the functional requirements of cooperation or on institutional design and effectiveness.[3] Such studies neglect to address the prior question of why some norms get selected over others, thereby defining international problems and guiding appropriate behavior in particular ways. Even studies that take a more overtly sociological approach have so far failed to adequately address this question. Whereas they provide mounting evidence that international norms and institutions may not only regulate behavior, but can also define state identities and interests,[4] few studies address the prior question of which norms get promoted or prevail over others in the first place.

This shortcoming is particularly evident in the literature on environmental institutions.[5] While research on the creation, design, and effectiveness of international institutions addresses crucially important questions, such studies generally lack a critical examination of what kind of governing norms institutions embody or why those norms came to dominate global environmental governance. Rather, an assumption often pervades the mainstream academic literature that any cooperation on environmental problems means progress toward a more ecological international order. A critical examination of the evolution of environmental norms shows that assumption to be overly simplistic, even faulty. Overcoming such lacunae in the literature deserves greater attention from scholars interested in the kind of international order that institutions actually promote.

In response, this study orients itself more toward what Robert Cox calls "critical theory." An exercise in critical theory need not invoke complex methodological or epistemological challenges to how scholars ought to go about understanding the world, a wholesale rejection of explanatory theory, or a radical interpretivism associated with some forms of postpositivist analysis. Rather, it simply poses the question differently than those involved in research on the important tasks listed above. As Cox puts it, "Critical theory stands back from the existing order of things to ask how that order came into being, how it may be changing, and how that change may be influenced or channeled. . . . Its aim is the understanding of structural change" (Cox 1992:3). In this spirit, I set out to answer two questions about international environmental governance: How did the current form of international environmental governance evolve since the first major international environmental conference in 1972? And, why did it evolve into liberal environmentalism while other alternatives fell by the wayside?

THE EVOLUTION OF INTERNATIONAL ENVIRONMENTAL GOVERNANCE

In the first half of the book, I detail the main empirical argument that norms of environmental protection have gradually converged with liberal economic norms in international environmental governance since 1972. The institutionalization of "sustainable development" at the 1992 UN Conference on Environment and Development (UNCED or Earth Summit) legitimated this convergence toward what I label liberal environmentalism.[6] This normative compromise predicates environmental protection on the promotion and maintenance of a liberal economic order. It also enabled environmental concerns to find such a prominent place on the international agenda.

Chapters 2 and 3 should not be read, however, as yet another exposition of what "sustainable development" means. Numerous works devoted to that topic only serve to highlight the ultimately elusive quest for a definitional consensus. The widely quoted definition in the Brundtland Commission report—"development that meets the needs of the present without compromising the ability of future generations to meet their own needs"—identifies a compromise between competing values including growth, conservation, and inter- and intra- generational equity.[7] As such, it is open to a myriad of interpretations. By the Earth Summit in 1992, various authors had found as many as forty definitions of the concept, and the Rio Declaration on Environment and Development, which articulated the global political consensus on sustainable development thinking, does not even attempt a consensus definition.[8] As noted by the International Institute for Applied Systems Analysis, the concept "is viewed quite differently by industrialists, economists, planners and environmental and ecological scientists . . ."[9]

Instead of unpacking the concept itself, I examine the international political and economic norms invoked in the name of sustainable development and trace their evolution. The focus on norms turns attention to the content of international governance. I define governance broadly as the methods or means of realizing shared values, interests, and goals that may or may not derive from a formal centralized political authority.[10]

At the basis of global governance are international norms, which define and regulate appropriate state (and other key actors') behavior, and assign rights and responsibilities regarding the issue in question. This definition corresponds to the constitutive, regulative, and deontic function of norms.[11] Norms constitute identities and meanings by defining who may act, in what context they may act, and what their actions mean in that particular context. They regulate by pre/proscribing how actors should behave in defined contexts (Cancian, 1975:5–7; Dessler 1989:456). Finally, norms serve a deontic function when they express values that create rights and responsibilities and thereby empower actors by providing reasons or justifications for particular actions (Onuf 1997; Ruggie 1998:21). All norms perform these functions simultaneously, but to varying degrees. In other words, international norms define, regulate, and empower legitimate state (and other key actors') behavior. While identifying institutionalized norms does not cover all aspects of governance, norms are at the heart of all governance structures.

When "sustainable development" appears as a goal in international environmental agreements, policy positions of multilateral agencies, or pronouncements of intergovernmental and even many nongovernmental fora, it evokes an identifiable set of norms that underlies recent attempts at international environmental governance. I detail the evolution of international environmental norms through three key points of norm articulation: the 1972 Stockholm conference, the 1987 World Commission on Environment and Development report, and the 1992 UNCED in Rio de Janeiro. I pay particular attention to how ideas developed that influenced the formulation of environmental governance through those episodes, who carried those ideas, and what form they took when they became institutionalized. Each event serves as a marker for what is actually an ongoing process of normative evolution.

Following an examination of each event, I describe the set of norms—or *norm-complex*—that represents the basis of environmental governance at that time. A norm-complex denotes a set of norms that governs relations of authority and the values promoted that define and regulate activities in a particular issue area. A norm-complex need not be stated explicitly, but can be inferred from specific norms. For example, John Ruggie (1998:62–84) identifies a norm-complex of "embedded liberalism" in the post-World War II era—where the liberal economic order is predicated on domestic intervention—based on specific norms constructed to govern international trade and finance as embodied in the Bretton Woods institutions. Similarly, liberal environmentalism can be inferred from the specific norms legitimated through the Earth Summit conference process and related agreements and activities. Like Ruggie, my purpose in articulating a norm-complex is to reveal the underlying pattern of values and goals that guide international behavior. Once identified, a norm-complex can be used to assess the significance of changes, which are best analyzed in relation to underlying *collective* purposes embodied in norms.[12] Threats to the norm-complex stem from alternative norms that undermine the collective purposes that define it. For example, a string of environmental agreements or programs that endorsed unilateral trade measures to protect the environment or denounced economic growth would constitute a serious challenge to liberal environmentalism.

Chapters 2 and 3 argue that the norm-complex governing global environmental practices evolved through stages that roughly correspond to the three events listed above. The norm-complex articulated at Stockholm in 1972 did contain both environment and development norms, but primarily emphasized the environmental protection side of the equation, as did the practices of international institutions, transnational activities, and international cooperative efforts following the conference. The Brundtland Commission report attempted a synthesis of environment and development agendas and reflected a Keynesian-like compromise. In it, liberal interdependence that generated growth would be tempered by managed interventions to cushion and facilitate adjustment in the South and direct development on a path less likely to harm the environment. Rather than call this norm-complex one of sustainable development, I label it "managed sustainable growth." I do so to contrast it with the currently dominant norm-complex of "liberal environmentalism" institutionalized at UNCED.

Liberal environmentalism accepts the liberalization of trade and finance as consistent with, and even necessary for, international environmental protection. It also promotes market and other economic mechanisms (such as tradeable pollution permit schemes[13] or the privatization of commons) over "command-and-control" methods (standards, bans, quotas, and so on) as the preferred method of environmental management. The concept of sustainable development, while it legitimated this shift in norms, now masks this compromise that characterizes international environmental governance.

A number of studies identify various elements of what I call liberal environmentalism, but they use terms such as ecological modernization or simply sustainable development.[14] Many of these studies aim to uncover contradictions in such concepts. Some also critique the form of environmental governance promoted as too accepting of the status quo of state control and of patterns of economic development and practices that created most of the world's environmental problems in the first place (see especially Chatterjee and Finger 1994). These critiques are not the focus of this study, although the concluding chapter addresses various implications of the institutionalization of liberal environmentalism. Instead, the value added here is to uncover how and why liberal environmentalism became institutionalized, at least at the international level, rather than simply offering a critique of the outcome. To date, no study has carefully traced through the institutionalization of such ideas. A critical understanding of the constraints and opportunities for change in international environmental governance requires understanding how that structure of governance developed over time.

EXPLAINING THE EVOLUTION OF ENVIRONMENTAL GOVERNANCE

Unlike studies that focus on how best to achieve international cooperation on environmental problems, in the second part of the book I explain why a particular set of norms dominates such arrangements. Explanations for the observed normative developments in environmental governance address two questions: What actors and institutions carried ideas that led to liberal environmentalism? And, why did the set of ideas associated with norms of liberal environmentalism become institutionalized, or, more generally, why does the set of ideas associated with a prevailing norm-complex become institutionalized over others?

The movement from one norm-complex to another occurs when a new set of ideas either redefines existing norms or introduces new norms not previously considered as relevant.[15] "Ideas" here simply refer to proposals for new norms, whether stated as an ideology, worldview, principled belief, or causal belief.[16] Each type of idea can be potentially stated as a norm if it provides an orientation to action. The importance of ideas in politics comes about through their collective legitimation—that is, when they take on a normative flavor—which is ignored by utilitarian theories focused on individual beliefs. The distinction between "idea" and "norm," although somewhat artificial,[17] distinguishes the initial articulation of ideas by individuals or groups, with the causal properties attributed to normative statements once they become a "collective intentionality" or institutionalized.[18]

I put forward two competing explanations for the selection of norms in chapters 4 and 5 that focus on the causal role of ideas in international governance. Such a focus is appropriate given the overriding concern with the content of international governance, since only knowledge or ideas-based explanations focus on what set of ideas dominate attempts to solve the problem in question.

Still, I take seriously Geoffrey Garrett and Barry Weingast's (1993:203) assertion that an ideas-based explanation first requires a null hypothesis that only material interests matter. Three difficulties, explored below, stand out that make standard power and interest approaches inadequate or inappropriate for this study. First, most such theories explicitly leave the content of interests unexplained. Second, even when state interests are known, outcomes in international environmental governance cannot be easily extrapolated from them. Finally, existing explanations that do link power and interests to the content of governance leave many of the most important empirical outcomes in environmental governance unexplained.

Limitations of Power and Interest-based Explanations

Leading power and interest-based approaches in International Relations do not, on their own, explain the normative content of institutions, first and foremost because interests are exogenous, thus unexplained in such theories. Robert Keohane forcefully makes this point in his discussion of neoliberal institutionalist and neorealist explanations for cooperation, the two dominant rationalist explanations in the literature:

> In the absence of a specification of interests . . . institutionalist predictions about cooperation are indeterminate.
>
> That is, institutional theory takes states' conceptions of their interests as exogenous: unexplained within the terms of the theory. Unlike naive versions of commercial or republican liberalism, institutionalist theory does not infer a utility function for states simply from the material economic interests or the alleged values common to democracy. . . . Nor does realism predict interest (1993:285).

As a result, theories that do attempt to explain normative outcomes almost always import interests, usually taking them for granted (Ruggie 1983:198). The literature on environmental institutions partakes in this bias, which accounts for its failure to explore adequately the collective purposes behind responses to global environmental problems.[19]

Analysts are then left with a very thin conception of interests. Depending on the theory of world politics employed, interests in international politics are assumed to be relatively stable and based on core values such as economic costs/benefits or protection of physical security from outside attack. In realist thought, interests are taken as given and the interests of dominant state-actors generally prevail, although the ability of those actors to prevail is conditioned on factors such as the current distribution of capabilities in the international system. Later institutionalist literature broadens the focus from the interaction of rational state actors to include institutional, transnational, and/or domestic factors to the mix of variables considered (Young 1994; Haas, Keohane and Levy 1993). Whatever the merit of specific rational-interest approaches to explaining cooperation (the critiques of specific theories are many and varied and need not be rehearsed here), by themselves they offer little in the way of analysis of the content of cooperative arrangements that prevail.

Instead, interest-based explanations for international phenomena have typically focused on the problem of how to achieve common interests or joint gains. Such explanations rely on modeling the strategic interaction of actors with given interests or they hypothesize that when core interests of powerful actors are threatened (for example, when they are vulnerable to costly environmental damage), those actors, either by threat, coercion, or by shouldering extra costs, will ensure action is taken in response to those threats. The outcomes explained by such theories are usually dichotomous— agreement/no agreement, action/no action, or cooperation/conflict—on the concern in question. Studies in this vein can provide interesting analysis of cooperation on particular international environmental issues. For example, Sprinz and Vaahtoranta (1994), while not explicitly presenting a theory of co-operation, use an interest-based explanation to show why some countries more strongly supported international environmental cooperation on con-trolling acid rain and ozone depletion than others. Paterson (1996) and Row-lands (1995a) assess the merits of interest-based explanations, among others, to explain the politics of global responses to climate change. These ap-proaches nonetheless fail to explain the construction of an "interest" in envi-ronmental protection, which would require a subsidiary theory of interest formation. More importantly, the actual framing or kind of responses to en-vironmental problems are also beyond the scope of such theories because a cooperative result, for example, says little about what goals or ends obtain in the cooperative solution.

Supporters of a more traditional liberalism that "takes preferences seri-ously" might still counter that the preferences of all relevant actors could have been gauged on the eve of the Earth Summit, and outcomes extrapo-lated from them (Moravcsik 1997). Such a unit-preference theory would propose that the South wanted primarily to develop and to defend sover-eignty, while it had a limited interest in the environment. Therefore, liberal environmentalism was the best that could be achieved given the North's desire to protect open trade and investment while at the same time appear-ing to do something about global environmental problems. Likewise, the North would have liked to see the South do more for the environment, but liberal environmentalism was the best it could do, given the South's inter-ests. Although this approach moves closer to addressing the problem of identifying interests, a simple extrapolation of outcomes from domestic preferences fails to explain the changes in the normative basis of environ-mental governance since 1972.

Earth Summit outcomes did indeed reflect, to varying degrees, the pur-suit of domestic material interests. However, if interests provided a suffi-cient explanation, those outcomes should have more closely resembled the competing sets of interests reflected in the Stockholm conference outcomes 20 years earlier, which for the most part simply juxtaposed environment and development. That UNCED produced different outcomes reflects in part that the South's "interest" in the environment changed drastically be-tween 1972 and 1992, as I show in chapters two and three. As one author describes it (Imber 1994:86), unlike the uniformly defensive position taken by the South at Stockholm, some countries in the South, as in the North, took positions at UNCED that could be considered "activist" on both environment and development concerns, others took "ambiguous" positions, and still others were on the "defensive." Indeed, the Stockholm outcomes appeared to reflect competing sets of material interests to a much greater degree since nascent ideas that linked environment and development had only limited influence. In the intervening years, new ideas that linked environment and development increasingly shaped interest-definition for countries in the North and South.

For example, positions changed substantially between 1972 and 1992 on arguments put forward by the South over the way in which development ought to be promoted within a framework that also considered environmental protection. The embrace (albeit grudging in many cases) of market norms by the South, the support for incorporating environmental concerns into development projects and policies, and the willingness to acknowledge the severity of global and local threats to the environment all represented substantial shifts from 20 years earlier. Likewise, the North's embrace of the concept of sustainable development in the late 1980s, with its explicit linkage of environment and development, cannot be derived from interests alone, unless those interests changed since 1972. The North resisted the linkage at that time because policymakers perceived the two interests to be fundamentally incompatible, and could not conceive of institutions that could promote both goals simultaneously. Admittedly, the priority given to particular environmental issues still often differed in North and South. Nonetheless, the shifts in positions noted above that facilitated the acceptance of liberal environmentalism suggests that its institutionalization reflects more than the sum of material interests or a simple North-South compromise. Ideas needed to intervene from some source to create, modify, or, at the least, find a focal point around which existing interests might converge and consensus might form.

Even in regard to the type of world order the North supported, nothing inherent in the material interest of a wealthy or powerful country makes it desire an economic system characterized by open trade and investment, the liberal side of liberal environmentalism. The Cold War period, wherein the two most powerful countries pursued vastly different conceptions of world economic order, not to mention the long periods of history where economic nationalist policies of powerful countries dominated, demonstrates that power and material interests do not dictate particular policy preferences in a predictable direction. Ultimately, even the quest to protect sovereignty on the part of the South cannot be considered solely a material interest since it is conditioned on a pre-existing set of social arrangements within the international system that privileges the role of sovereign states and defines relations between sovereign states in particular ways.

Finally, even if a focus on domestic preferences could explain outcomes at given times, such explanations remain extremely inefficient, requiring constant reevaluations of changing preferences over time, and would still require a domestic theory of preference formation and change. An explanation that either attempts to explain why interests changed or that endogenizes interests to some degree would be more efficient. Chapter 5, as I explain below, puts forward a more efficient theory than a pure rational-interest approach because it endogenizes the evolving normative context of state practices rather than relying on either the repeated evaluation of particular state preferences or the uncritical importation of assumed interests.

A third possible approach is to employ theories that attempt a more explicit linkage of power and interest to norms, most notably those that focus on hegemony. They nonetheless still encounter the theoretical pitfalls of standard rational-interest approaches. Moreover, they have performed poorly as explanations of environmental governance in studies to date and the evidence in subsequent chapters lends little support for such an explanation in this case.

The most prominent example is Hegemonic Stability Theory, developed in the 1980s. In its various forms, the theory explains which norms prevail by arguing that international regimes, and the norms they embody, reflect the interest of a dominant or "hegemonic" state in the system, or, in some modified versions, a group of dominant states (e.g., Snidal 1985). This theory served as the fallback position to explain the creation of international institutions for much of the "international regimes" scholarship, and for questions of world order more broadly. For example, Robert Gilpin, Robert Keohane, and Stephen Krasner—despite differences in their work in other respects—all implicitly or explicitly supported the view that a hegemonic state that is able and willing to play a leadership role, is necessary (though perhaps not sufficient) for the establishment of international regimes.[20]

This state-centric view of hegemonic stability theory has proven of limited usefulness in explaining either international environmental cooperation or normative development. Oran Young, for example, has shown in a series of articles that hegemonic leadership was not necessary for the formation of many multilateral environmental agreements and sometimes played only a minor role, even when a hegemonic power participated in the regime or agreement eventually (Young 1989, 1994). Even if one considers the United States a hegemon in the environmental issue area (a dubious proposition in any case) the experience of the Stockholm and Rio conferences seems at odds with the basic hypothesis of hegemonic stability theory.[21] At Stockholm, the United States did play a leadership role, although the agreements reached did not reflect U.S. interests solely, especially since it showed very little interest in the development side of the agenda. At Rio, the United States was a disengaged player for much of the negotiations, while the European Union (EU) assumed a much more assertive leadership role (Hajost 1994; Sjöstedt et al. 1994). However, in both cases, the basic framing of issues and interest-definition came as much from the entrepreneurial leadership of the conference secretariat as from particular states, and drew from ideas and institutional developments not directly derived from dominant state interests.

Granted, dominant states may be able to block agreement on or effectively veto international norms, since they may provide the resources needed to implement the norm or their practices may be a vital part of those that the norm targets. For example, the norm of additionality—that aid transfers for environmental matters from North to South ought to be new and in addition to existing transfers—probably owes its lack of successful institutionalization to consistent opposition by the United States.[22] Nonetheless, little evidence supports the position that given interests of dominant states determine what norms *will* actually arise. One would be hard pressed to make the argument, for example, that norms of environmental protection could be derived from the structural power position of major states, (which requires a questionable theoretical strategy in any case as noted above). The role of interests then cannot be easily derived from a material structural theory of international politics. As will be shown in subsequent chapters, the pursuit of interests by states and groups of states in particular negotiations occurred within a broader normative context that shaped those interests, even those of dominant states.

A more classical argument on the same theme focuses not on state power *per se*, but on ideational hegemony within particular world orders. This Gramscian approach argues that both the dominant state's interest and the international order rest on the ideational hegemony of a dominant class privileged by the current global mode and relations of production. To simplify a varied literature, the central argument of Gramscian and historical materialist approaches is that the international order serves the interest of the class privileged by the international division of labor which reflects, in the modern world, the current stage of capitalist development and economic relations (Gill 1993; Cox 1983, 1987). International organizations and regimes thus reflect such class interests, and give legitimacy to these interests.[23]

While these approaches pay more attention to interest-definition and ideational factors, they still say little directly about an issue area such as the environment, interest in which cannot be derived strictly from economic structural factors. Their main advantage is to open up critical appraisals of prevailing practices by shifting the focus from multilateral cooperation to the underlying structural conditions that give rise to environmental degradation. They can also reveal contradictions in environmental policies (and the potential of such contradictions to produce historical change) and the underlying patterns of capitalist production that may (or may not) contribute to environmentally destructive patterns of development. These radical critiques, however, while revealing of evidence obscured by rational-interest approaches, offer more in terms of description than explanation.

Gramscian approaches are weakest, however, in explaining the dynamic processes through which responses to environmental problems are shaped or why the environment has become a mainstream issue in international politics at all, except by post hoc reasoning (Williams 1996:51–52). A Gramscian analysis is consistent with some of the patterns of governance identified in chapters 2 and 3 since liberal environmentalism could certainly be viewed as a way to legitimate or provide optimal political and economic conditions for the maximization of private capital returns (by supporting, for example, the free movement of goods, capital and technology) while appearing to respond to growing environmental concerns. But much remains vague and unexplained. The overly general treatment of forces of capitalist production behind a liberal economic order, and underspecification of the links between material forces and class interests, and the institutions that promote, sustain, and legitimate hegemonic orders, limits the ability of Gramscian approaches to explain how environmental governance has evolved. Particularly obscure is the role of agency in promoting new values such as environmental concern, since, evidence suggests, these concerns did not originate from the interests of capital.

For example, the resistance to liberal environmentalism by segments of industry suggests that ideas played an independent role from the structural dictates of capital (or else globally minded capitalists did not act in their class interest). Those who suggest that business did play a privileged role in the support of what I call liberal environmentalism point to the close relationship between UNCED secretary-general Maurice Strong and Stephan

Schmidheiny, who founded the Business Council for Sustainable Development (BCSD) with Strong's "encouragement and support."[24] Strong appointed Schmidheiny as one of his top advisers and gave funding to BCSD in the lead-up to Rio.[25] The group had representation from a number of the largest multinational corporations in the world and its popular report, *Changing Course*, articulated a version of environmentalism that meshed closely with many of the Rio outcomes (Schmidheiny 1992). However, the ideas contained in *Changing Course* were already well established within organizations such as the Organization for Economic Cooperation and Development (OECD) and the EU. While Gramscians might be correct that such institutions often function to further the legitimation of hegemonic orders, they are also forums where new ideas can be generated that redefine, in keeping with this language, the form that hegemony will take. Just as an analysis of modern capitalism requires explanations of the rise and decline of the Keynesian welfare state that look to ideas as well as economic variables, so too does change in global governance, where international institutions sometimes operate like the state in domestic society, that is, as at least somewhat independent of as well as interactive with the market and the power of private capital.

In the case of environmental governance, whereas the privileged access given to business, or at least Schmidheiny's group and perhaps the International Chamber of Commerce at UNCED is beyond doubt, industry as a group cannot be credited with formulating ideas around sustainable development or with being overly significant players in its institutionalization, although BCSD's support surely helped the profile of UNCED in the international business community. As subsequent chapters demonstrate, industry consistently came late and often fought the compromises that eventually evolved into liberal environmentalism.

A key example is the resistance of a major industry lobby to findings of an OECD conference in 1984 on Environment and Economics (OECD 1985). As chapter 5 will show, this conference articulated many of the core ideas that would later inform the Brundtland Commission report. Among the varied nongovernmental groups and members of governments involved,[26] only the Business and Industry Advisory Committee to the OECD issued a separate statement that qualified the findings of the conference (OECD 1985:243–247). The group showed resistance to the idea that the economy and environment can be mutually reinforcing, a key finding of the conference and the Brundtland Commission. Instead, it argued, "there should be a balance between environment policy and economic policy," indicating that it viewed the balance of environmental and economic policies as a zero-sum game. Furthermore, industry has tended to be reactive rather than proactive, and in general has shown resistance to environmental policies when they threaten particular interests of individual industries or sectors. This approach outweighs the limited attempts on the part of industry to fit policies into an overall structure supportive of liberal norms.

In addition, little unity can be discerned among industry groups on shaping environmental norms in the main period under investigation. Rather, corporations tend to address specific issues based on how policies directly affect their profits. For example, in negotiations on a climate change convention, oil producers—who formed their own nongovernmental organizations such as the Global Climate Coalition (GCC)—actively lobbied to prevent any regulation or action that might limit oil consumption.[27] Meanwhile, renewable energy providers, through the World Sustainable Energy Coalition, emerged as supporters of greenhouse gas emission reductions. Similarly, an alliance arose in the mid-1990s between the environmental group Greenpeace and the insurance industry, which feels vulnerable to catastrophic weather events that may come with global warming (Paterson 1996:164–167; Rogers 1993:244–245).

Finally, observers and analysts of environmental policies have noted that while corporate head offices have endorsed cost-effective, market-oriented approaches to environmental protection for some time, industry lobbyists have shown "a curious resistance" (Hahn and Stavins 1991:25). Hahn and Stavins, for example, note this trend in a study of the switch from command-and-control to market-based policies. Specific businesses or industries may resist, they argue, because although market-based policies may provide a given level of environmental protection at minimum cost for society as a whole, they often involve substantial transfers between sectors. Thus the changes I identify toward liberal environmentalism appear to have pulled industry along, rather than vice-versa.

In sum, a Gramscian argument ultimately rests on an overly blunt explanatory scheme where classes empowered by the current mode of global production ultimately triumph. If those classes are in fact pulled along, then other causal factors must also be accounted for in any explanation of what international order prevails. A broader historicism that takes account of intellectual movements and noneconomic social forces would open up analyses of other causal linkages, but then also moves beyond variables identified as most important by neo-Gramscian contributions to international relations.[28] My proposal below for a "socio-evolutionary" explanation builds on the neo-Gramscian insight that ideas interact with broader international structures, but attempts a more generalizable explanatory scheme by focusing on international social structure more broadly rather than the structure of capitalist production or the implicit assumption that ideas that succeed are necessarily generated to serve the interests of capital.

The inadequacies of strictly rationalist, power and interest-based explanations point to the need for alternatives that examine the causal role of ideas in international relations.

Ideational Explanations

The recent attention to the role of ideas in international politics arose to overcome limitations of rationalist approaches by including explanations of why a reconceptualization of interests occurs. The literature is split between those who treat ideas as intervening variables between interests and behavioral outcomes and those who take an interpretivist approach that focuses on the persuasive power of ideas or their embodiment in discourses.

The former approaches have the advantage of easily fitting into a positivist epistemology where outcomes can be clearly coded and the conditions for those outcomes to occur may be identified in a testable way. Goldstein and Keohane (1993a), for example, take a rationalist approach, in which ideas inform the preferences actors bring with them to strategic interactions, they provide focal points for cooperation when obvious equilibria are absent, or they act like Max Weber's famous "switchmen," directing future policies along certain paths. In all these cases, they view ideas as important not because of their meaning, but because they provide solutions to rational cooperation problems or because they are functional for institutional stability. Nonetheless, they admit their approach does "not suggest a theory for the creation of [ideational] switches, or even a fully worked-out model to explain the process by which ideas are selected" (Goldstein and Keohane 1993b:12). The causal capacity of ideas—their ability to provide reasons for actions based in their meanings—is beyond the scope of their approach (Ruggie 1998, 22; Woods 1995; Yee 1996).

To achieve a more direct focus on the content, or ideational basis, of environmental governance structures, other authors have turned to discourse-theoretical approaches. For example, Maarten Hajer (1995) has used discourse analysis to examine how the discursive practices around "ecological modernization"—the notion that environmental problems can be solved in accordance with the workings of the main institutional arrangements of society—influenced the regulation of environmental conflict around acid rain policy in the United Kingdom and the Netherlands. Similarly, Karen Litfin (1994) has shown how the framing and interpretation of scientific knowledge shaped international responses to ozone depletion. A discourse approach performs best as a way to understand social context and meanings that constrain and enable certain policy choices and regulate social conflict. It can generate revealing descriptions that identify changing discourses and how such changes might influence the legitimacy of particular policy choices.

Discourse analysis performs less well in specifying actual practices or institutional arrangements that prevail. Discourse is significant when it becomes institutionalized, as competing discourses are constantly present. But discourse analysis alone occurs almost exclusively in the realm of language and meaning, thus is weak in its analysis of political processes that make some discourses heard over others. This method of analysis also makes it difficult to separate the sources of discourse from its effects and runs into the danger of presenting discourses as if they float freely,[29] to be snared by actors in a fashion left unexplained within such approaches.[30] Discursive approaches, while equipped to analyze symbolic languages and intersubjective meanings, thus face criticism for not being sufficiently explanatory because they emphasize "the interpretation of meaning and/or the ambiguity and instability of all interpretations" (Yee 1996:200; Blyth 1997).

The attempt here is to ground discussion in how, when, and why ideas became institutionalized. I therefore avoid the ambiguities of a focus on the role of discourse and instead examine agreed-to norms and where they originated. The nature of norms as based in intersubjective meanings suggests that any norm-based analysis, including this one, shows an affinity to what social theorists refer to as discourse. However, this study attempts to ground its analysis more in practice and institutionalization as these have identifiable empirical referents. The focus below on the movement of ideas to norms accomplishes this task.

From Ideas to Institutionalized Norms: Epistemic Communities and Socio-evolution

Chapters 4 and 5 test in detail two approaches that focus on how ideas associated with liberal environmentalism became institutionalized. These approaches respond more directly than those just reviewed to the two questions posed at the start of the section—who carried ideas associated with liberal environmentalism and why did they become institutionalized?

Chapter 4 tests an "epistemic communities" explanation that looks primarily at actors who carry or create ideas. It argues that the ability of new ideas to become institutionalized rests primarily on the legitimacy of their source, focusing especially on the role of groups of scientific experts. It asserts *that scientific consensus within an epistemic community, "politically empowered through its claims to exercise authoritative knowledge and motivated by shared causal and principled beliefs," and its promotion of norms derived from that consensual knowledge, leads to the adoption of its ideas over others as guides to appropriate behavior* (Haas 1992a:41—emphasis mine). Following an argument put forward by Peter Haas in the environmental issue area, I test the claim that new norms arose in response to consensus within a "scientific ecology" community of experts (Haas 1989, 1990, 1992a, 1992b, 1996; Haas and Haas 1995). In undertaking a thorough analysis of the influence of scientists and scientific ideas on the evolution of environmental governance, chapter 4 not only tests an epistemic communities hypothesis, but also offers more general insights into how science and politics have in fact interacted in the shaping of environmental governance.

The epistemic communities explanation has been used primarily to show how consensual knowledge within such groups aids international policy coordination by redefining state interests to facilitate rational cooperation. However, I am drawn to it here more for its underlying assertions about how and why a particular set of ideas (in this case, those associated with an expert group of ecological scientists) comes to dominate cooperative outcomes. An epistemic communities approach thus offers clear answers to questions of both who carried ideas (scientists) and why those ideas had causal weight (legitimacy of their knowledge claims).

Despite recent critiques, the epistemic communities literature still provides the clearest explanatory framework available of how scientific knowledge translates into changed patterns of state behavior and international interactions (Susskind 1994; Yee 1996). It has also been influential in scholarly work on the role of ideas in international relations more broadly (Goldstein and Keohane 1993b:11 fn. 18; Yee 1996; Litfin 1994). Furthermore, it often serves as a point of departure for studies on international environmental action since the complex and uncertain nature of environmental problems appears to privilege experts in determining the nature of environmental problems and the technical requirements needed to address them. These factors combine to make environmental governance a paradigmatic or crucial case for the approach in that it should perform best in issue areas characterized by uncertainty and technical complexity.

Ultimately, I find that an epistemic communities approach fails to account for normative evolution in this case. Chapter 4 demonstrates that consensus on both cause-effect knowledge and values within the relevant communities of experts were weaker than often portrayed. Furthermore, core ideas of environmental governance did not originate from a single identifiable epistemic community, and often contradicted the preferred outcomes of scientific ecology, as Haas defines it. Indeed, the historical evidence suggests that the causal arrow often ran in the opposite direction, with ideas around liberal environmentalism increasingly influencing global environmental research.

Despite this finding, chapters 4 and 5 should be read as more of a rescue than a wrecking operation. They highlight that important insights underlying the epistemic communities argument can be too easily dismissed when tied to the dominant way in which the epistemic communities argument has been presented in empirical research. First, its basic insight about the importance of knowledge and discourse in global policy is too easily dismissed because even friendly critics find that studies that apply the argument to explain policy change or coordination overplay the causal significance of particular expert groups (agency) and their influence through domestic bureaucracies. Second, perhaps for strategic reasons on the part of its proponents, most published research in this vein remains wedded to a rational institutionalist research program focused on the problem of co-operation rather than broader questions of political change.

Chapter 5 can be read as an attempt to recapture many of the core insights of the research program (e.g., Adler and Haas 1992), while also moving it in a new direction with greater sensitivity to wider constraints of international social structure. It introduces an alternative explanation that focuses on the causal role of economic ideas, but not simply as embodied in an epistemic community of economists. Instead, I put forward a *socio-evolutionary* explanation that draws from recent work on the evolution of norms in international politics.[31]

This explanation begins not with actors or state power and interests (as do liberal and realist explanations), nor with economic structures and class interests (as do Gramscians) but with systemic social structure. It contends that explaining the selection[32] of norms requires an examination of the interaction of ideas (proposals for new norms) with the social structural environment of institutionalized norms they encounter. The main argument is that the *social* fitness of proposals for new norms with extant social structure[33] better explains why some norms are selected, while others fall by the wayside.[34] This starting point follows from insights from "constructivist" scholars of international relations who argue that interests themselves are derived, at least in part, from an existing social structure of norms and institutions in which actors participate (Finnemore 1996a, Katzenstein 1996, Ruggie 1998).

The explanation argues that three factors determine the selection of new norms: *the perceived legitimacy of the source of new ideas; fitness with extant international social structure; and fitness with key actors' identities at various levels of social structure.* By identities I mean both their status as agents as constituted by international social structure and their socioeconomic identities generated domestically which they project in their international affairs, such as their view of legitimate political and economic order as reflected in domestic institutions.

Since social structure and state identities and interests are mutually constitutive, this explanation does not exclude material interests or power as important factors in the selection process. Rather, by using extant social structure as a starting point of analysis it endogenizes an important source of interests, and thus offers a more efficient explanation. In so doing, it takes a modest step toward responding to James Caporaso's challenge to constructivists, "to explain institutions and sociality given some data on extant and prior institutions and sociality" (1993:82–83).

Chapter 5 goes over the socio-evolutionary approach and concepts used in detail. This explanation reveals the importance of particular sets of economic ideas, but attempts to push the current ideas literature further by emphasizing the interaction of new ideas with an existing social structure. In this case, the approach highlights how a group of policy entrepreneurs, drawing primarily on a set of economic ideas, were able to successfully reframe norms of international environmental governance to fit better with the broader international social structure than had previous attempts. The resulting compromise of liberal environmentalism made possible increased international efforts to address environmental problems and shaped how responses would be framed.

The concluding chapter discusses some of the implications of my findings for international environmental governance and for explaining and understanding change in international or global governance more broadly. It suggests that certain kinds of knowledge and policy responses are privileged not because of their inherent truth or even effectiveness, but because the institutionalization of liberal environmentalism grants them legitimacy. The legitimation of liberal environmentalism in turn has and will open up and close off various courses of action, with important implications for our ability to manage global environmental problems. The implications of these constraints on current and future policies and new potential sources of change are also explored. Knowing the origins of these norms and the processes through which they become institutionalized contributes to opening up critical questions of the kind of order institutions promote, rather than taking the progress of international environmental cooperation for granted.

METATHEORETICAL ISSUES

The socio-evolutionary approach and the focus on norms more generally raise epistemological issues of explanation and causality that I want to address at the outset, since they may be a source of confusion.

The socio-evolutionary approach developed in chapter 5 is an attempt to move the discussion away from a focus on an expert group alone, and toward the interaction of ideas with their environment. In the formation of international norms, that environment is the existing set of institutionalized norms that make up international social structure. Because international social structure is constantly evolving in response to the institutionalization of new norms and altering of old ones, the socio-evolutionary approach lends itself naturally to a historical and interpretivist methodology. The content, in terms of meaning, of social structure must be investigated at any given time as the environment in which new ideas compete. The approach is limited in its ability to model mechanical causal relationships because the fitness of ideas associated with liberal environmentalism is historically contingent.

What kind of explanation, then, does a socio-evolutionary approach entail? Two kinds of explanation are actually at work in the socio-evolutionary approach: causal and constitutive (Wendt 1998). The explanation is causal because it identifies factors that make some ideas more likely candidates for institutionalization or legitimation than others. Even though these factors, such as fitness with social structure or promotion of ideas through legitimating institutions, are based in part on intersubjective understandings of meaning, they can still possess causal weight, and have empirical referents. However, the focus on social structure means a constitutive explanation is also at work (although social structure also produces causal effects). Constitutive explanations answer "how-possible" or "what" questions. The goal of a constitutive explanation is to "account for the properties of things by reference to the structures in virtue of which they exist" (Wendt 1998:105). Social structures have constitutive effects that explain how something (such as an event, practice, or relationship) is possible, or what its properties are.

Indeed, the outcome to be explained here is not cooperation, but the very meaning and understanding of the global environment, which is part and parcel of the constitution of global environmental governance. The two are inseparable in at least one explanatory sense: international environmental governance makes sense only when relevant actors understand who is part of the international and what is being governed. States, sovereignty, property rights, global commons, pollution, and so on are all socially constructed, but have constitutive and causal consequences for what international environmental governance is and what actors engaging in the process are likely to do.

Because both causal and constitutive explanations *are* explanations (not merely descriptions), the approach does not entail a complete rejection of neopositivist understandings of evidence or testability. Thus, unlike postmodernists, poststructuralists and other forms of postpositivism, the socioevolutionary explanation argues that the context for action (social structure) can at any given moment in time be held constant for the purposes of analysis, and has a determinative content (norms and institutions) which can be gleaned through careful historical analysis and informed interpretation. Although this explanation differs ontologically from dominant positivist perspectives in International Relations in stressing intersubjectivity, epistemologically, it falls into the "modernist" school of constructivist scholarship, which does not preclude the use of "standard (positivist) methods alongside interpretive methods" (Adler 1997:335).

A second source of confusion arises with the terminology of causation itself. This confusion can be addressed by contrasting the notion of causality employed here with a mechanistic or Newtonian view of causation, which is still prevalent in international relations research (e.g., King et al. 1994). That version of positivist epistemology, with its roots in Humean empiricism, makes no claim about actual forces of causation. Rather, it seeks to identify regular and predictable series of events that occur whenever the same conditions hold. Causation, in this view, exists outside of time and space. The positivist method is to propose a generalization or theory about some event or situation, deduce a testable hypothesis from the theory, and observe whether the prediction succeeds. Positivist theory is predictive to the degree that one gains confidence, through testing (of falsifiable hypotheses) and modifying of theories (or generalizations), that one has identified necessary and sufficient conditions, which can then be inferred to cause the outcome in question.[35]

However, the social world, including international politics, does exist in time and space, and thus is indeterminate. John Ruggie, beginning with this observation, contrasts the Humean notion of causality above with what he calls "narrative" causality. This notion of causality "conforms to its ordinary-language meaning: whatever antecedent conditions, events, or actions are significant in producing or influencing an effect, result, or consequence."[36] This difference is important for studying intentionalistic and reflective human beings in history, where generalizable mechanical laws do not always apply.[37] The social world consists of reasoned and intentioned action as well as unintended consequences, the causes of which can be found as much in ideas, norms, and institutions as in the physicalist universe of "distinct actors, with palpable properties, engaged in discrete events" (Ruggie 1995:96). As already mentioned, international relations theories that focus solely on power and interest—which can most easily be modeled in the physicalist universe of positivist epistemology—may tell us something about the form of international relations (whether cooperative outcomes are more or less likely, for example) but little about the content.

A focus on content also means turning to "social facts" of intention and meaning.[38] "Social facts" are not "causes" in a simple mechanical way via external constraint, but can internally or cognitively define and redefine the identities and interests of actors. Constructivist international relations theory has made much of this aspect of international norms and institutions, which includes both the idea of constitutive or enabling rules as opposed to the strict focus on only specialized regulative and enforcement rules that characterizes rational-interest approaches to norms and ideas, as well as the world of intentions and meanings.[39] Norms condition the possibilities of action by defining the range of meaningful if not of conceivable behavior, and by pre/proscribing the types of appropriate or legitimate behavior that can be performed in particular social contexts. The institutionalization of norms has causal effects because it increases the likelihood of the behavior they prescribe and decreases the likelihood of the behavior they proscribe. While this may not guarantee that all behavior will conform to the norm, it shifts the burden of effort and proof onto those actors who contest its validity, and empowers actors in conformity with the norm. Following such reasoning, Yee (1996:97) argues that norms "quasi-causally affect certain actions not by directly or inevitably determining them but rather by rendering these actions plausible or implausible, acceptable or unacceptable, conceivable or inconceivable, respectable or disreputable, etc." Unless one is confident that knowledge about international politics can be derived solely from the brute facts of power politics, a posi-

tion I have already shown to be inadequate, then one must be open to a notion of causality that recognizes the causal power of human intentions and
reasons (ideas) and the norms and institutions (social structure) that provide the intersubjective context of human action.

Building explanations with a notion of narrative causality in mind has
two advantages according to Ruggie. First, it focuses on linking events to
one another over time to discover their effects, even when such "events" include thickly described social facts. The second is that it organizes these
statements into an "interpretive 'gestalt' or 'coherence structure.' "[40] This
method, which Polkinghorne calls "emplotment" is "not the imposition of
a ready-made plot structure on an independent set of events; instead it is a
dialectic process that takes place between the events themselves and a
theme which discloses their significance and allows them to be grasped together as parts of one story" (Ruggie 1998:94). The goal, according to Ruggie (1998:94), "is to produce results that are verisimilar and believable to
others looking over the same events," not to produce a covering law across
time and space.[41]

Following on these metatheoretical positions, the socio-evolutionary
approach is explanatory because it not only identifies social structure and
posits its explanatory power in the constitutive sense, but also emphasizes
that specific factors can be identified that reveal processes through which
intersubjective meanings evolve. In chapter 5 I will identify the general
contours of social structure that provides the environment with which new
ideas interact, a step glossed over in the often loose formulations of recent
theoretical work in this vein.

The question of prediction also arises whenever one makes causal
claims. If some force, social fact, or material condition possesses causal
weight, its presence or absence would be a basis on which to expect certain
outcomes. The difficulty again comes with equating narrative causality
with the more mechanistic, formal causality. The latter implies a precision
in predicting outcomes given the presence or absence of particular causes.
The former recognizes the indeterminacy of human action, and the contingent nature of the conditions it views as causes. Liberal environmentalism is not a dependent variable that can be measured along a quantifiable
axis. Rather, it is an intersubjective understanding of international environmental governance that exists within a particular historical context.
The causes are thus also historically specific, and their significance can only
be determined through careful historical and interpretive analysis of the
ideas that led to those understandings and the processes through which
those ideas were selected. Counterfactuals are one useful methodological

tool to evaluate the causal role of ideas or of the norms they informed. One might ask, for example, what would have happened in the absence of ideas associated with liberal environmentalism? This question is entertained, for example, in chapter 5, where it is noted that the idea of ecodevelopment would have led global environmentalism in a different direction if alternatives along the lines of liberal environmentalism had not come along.

In addition, the success of some ideas can be compared to the failure of others to look for clues as to the selection mechanisms at work. Thus, all is not contingent under this explanatory framework. Given an extant social structure, even if historically contingent (i.e., the social structure itself also evolves over time), selection processes determine or explain ideational success. The socio-evolutionary approach in chapter 5 does identify a selection process that makes it more likely that some ideas will be selected over others in becoming institutionalized, and in that sense it is predictive. But given the historical contingency of social structure, and even the generation of new ideas, the explanation cannot be fully predictive since the conditions that hold within a given social structure may not hold in the future. A socio-evolutionary explanation is not predictive in the sense of specifying generalizable antecedent conditions for particular outcomes. Neither is the theory predictive in a functional sense, that is, based on a logic—such as reproductive logic in evolutionary theory—that explains normative outcomes simply by the benefits social structure provides for particular norms.[42] Thus, although evolutionary terminology is used in both cases, the view of history is quite different than that type of rational evolutionary argument common in the economic literature on institutions. That approach views the emergence and change of institutions as efficient responses to their environment. History is simply an efficient response to changing technology or other material factors in the external environment, where the most efficient (that is, "fit") institutions consistently win out. A method of comparative statics suffices over narrative (Caporaso 1993:79). In contrast, the socio-evolutionary approach views history similarly to sociological institutionalists, who, as Caporaso explains, stress the "contingent, path-dependent nature of institutional change." Choices made by individuals cannot be explained in purely instrumental terms because "their future choices, their perceptions of what is possible, and their beliefs and standard operating procedures are products of past historical choices."

In the case of socio-evolution, the explanation is historical, in that the mechanism of change I focus on stems from conscious efforts at social construction, which have no parallel in evolutionary theory. The selection process that occurs through the interaction of these efforts with social structure does not determine these formulations, but selects them through an interactive, recursive, and historical process, as agents are often aware of the social structural reality within which they interact, even if some structures are so deeply institutionalized that they are taken for granted. Given the contingent and historical nature of social structure, narrative methodology drives out comparative statics.

The place to begin such an analysis, then, is with the story of environmental governance as it evolved over the last thirty years.

Chapter 2
FROM ENVIRONMENTAL PROTECTION TO
SUSTAINABLE DEVELOPMENT

TO ASSESS WHY certain norms prevail, the first task is to identify the set of institutionalized norms—or norm-complex—that defines and regulates appropriate behavior, and assigns rights and responsibilities regarding the issue in question. This chapter and the next undertake this task in detail, a step often omitted in institutional analyses of environmental governance. The two chapters are organized around the major defining events in international environmental governance over the last thirty years: the 1972 United Nations Conference on the Human Environment (UNCHE) in Stockholm; the 1987 World Commission on Environment and Development (WCED) report *Our Common Future* (also known as the Brundtland Commission report); and the 1992 United Nations Conference on Environment and Development (UNCED) in Rio de Janeiro.

This chapter tells the story of the politics and outcomes of the first two events, and traces the development of ideas on environmental governance that occurred leading up to each event and in the intervening period. The trace of norms begins with a relatively detailed account of the Stockholm conference because the origins of the compromise of liberal environmentalism can be found there. At Stockholm, the concerns of conservationists and environmental scientists (primarily from Northern countries) encountered resistance from states more concerned with economic growth and poverty reduction. From this confrontation, new thinking developed that attempted to link environment and development into a single framework under the rubric of "sustainable development."

Chapter 3 picks up the story following the publication of the Brundtland report. The report marked the first real synthesis of the environment and development agendas and reflected a Keynesian-like compromise. In it, liberal interdependence that generated growth would be tempered by managed interventions to cushion and facilitate adjustment in the South and direct development on a path less likely to harm the environment. By 1992 a shift in norms of environmental governance had occurred, characterized by a general acceptance of liberalization in trade and finance as consistent with, and even necessary for, international environmental protection. These norms also promoted market and other economic mechanisms (tradeable pollution permits, privatization of the commons, and so on) over strict regulations (standards, bans, and quotas) as the preferred method of environmental management. The Earth Summit institutionalized this set of norms, which, the chapter will show, continues to guide what are viewed as appropriate responses to global environmental problems.

In both chapters, lists of international norms follow the sections on these three pivotal events. The lists illustrate the evolution from a primary focus on environmental protection to the current norm-complex of liberal environmentalism.

Before launching into the detailed tracing of this normative evolution, I build on my earlier discussion of what norms are in order to clarify how I identify which norms prevailed during the periods under investigation.

IDENTIFYING NORMS

Following from the definition given in the introduction, norms do not necessarily identify actual behavior; rather they identify notions of what appropriate behavior ought to be. Whereas a growing body of international relations scholarship emphasizes the intersubjective, or shared, nature of international norms, this is true only in the sense that they are irreducible to individual beliefs. What makes a norm "collective" is its *institutionalization*, which concerns the perceived legitimacy of the norm as embodied in law, institutions, or public discourse even if all relevant actors do not accept it or follow it (Onuf 1997:17; Jepperson, Wendt, and Katzentein 1996:54, fn 69; Busumtwi-Sam and Bernstein 1997).

Legitimacy matters because the question is not whether the norm exists, but the political authority the norm enjoys. Institutionalized norms constitute social structure and thus define which political institutions and practices are viewed as appropriate. A claim of legitimacy does not necessarily mean it adheres to a deeper notion of justice. Rather, norms are legitimated externally through political processes; they obligate because of agreement of members of the relevant community (Florini 1996:364–365; Franck 1990:16, 38). The degree of institutionalization is important because it indicates how durable the norm is likely to be, how strongly challenges to it are likely to be contested, and ultimately the ability of the norm to (re)define state interests.

Being collectively held, norms are "discrete positivities" and thus can be operationalized more straightforwardly than often portrayed (Onuf 1997:32; Raymond 1997:219–222). Most international norms are stated explicitly in treaties and conventions, less formal agreements, rules and standards established by international organizations, resolutions, and declarations, including the "soft" declaratory law that has served as a basis for international environmental law and institutions (Chinkin 1989; Dupuy 1991). Uncodified norms may be inferred from these same sources plus judicial decisions, statements by leaders or from state practices in given interactions. In this case, the relevant laws and organizations investigated include declarations and treaties from major environmental conferences, and policies and practices of organizations such as the United Nations Environment Programme (UNEP) and the World Bank, as well as statements, policies, and practices of relevant states. Hence, the empirical outcomes on which I focus are not mere exhortations, but leave behavioral traces and verifiable evidence in the form of treaty commitments, action programs, policies and policy instruments, and so on.

The degree of institutionalization can be inferred primarily from the norm's frequency or "density" in social structure, that is, the amount and range of instruments, statements, and so on, that invoke the norm (Florini 1996; Krasner 1988), and the degree to which actions that violate the norm require justification.[1] Violations often bring a norm into sharp relief because they either require justification (reinforcing the norm) or bring forth denunciations or attempts to replace the norm with an alternative (undermining the norm's legitimacy) (Finnemore 1996b:158–159). Thus, noncompliance alone does not undermine a norm's legitimacy (Franck 1990:151). Although repeated violations of a norm undermine its legitimacy, more important is when prominent actors denounce the norm or attempt to replace it with a competing one. This corresponds to the practice in international law of inferring custom from the consent of states or failure to "persistently object."

Following these criteria to identify norms, I analyze the outcomes (treaties, declarations, action plans, and so on) of the three major United Nations environmental initiatives listed above to determine the pattern of normative evolution. Scholars point to these initiatives as key turning points that generated political legitimacy for sets of norms that shaped appropriate responses to environmental problems that followed them. They articulated, more than any other events, the consensus (or conflicts) on norms at those times.

The lists of norms that follow the sections on the three events are based on the analysis below of outcomes, related agreements, and scholarly assessments of the lasting influence of the initiatives. Each list classifies norms in three broad categories that correspond to the nature of the actors the norms empower and their rights and responsibilities; the political economy of the issue; and the specific management norms promoted. The categories were chosen mainly for comparison purposes, being relevant to the environmental issue area and because they highlight the main fault lines of contestation. The categories also allow the analytic distinction between norms that constitute basic actor identities and norms that define and regulate the economic and political relationships between those actors. Table 1, which follows the discussion of UNCED norms in chapter 3, summarizes these findings.

UNITED NATIONS CONFERENCE ON THE HUMAN ENVIRONMENT (UNCHE)[2]

The Stockholm conference's significance lies in its articulation of a nascent set of norms that would become the basis for international environmental law and practice (Schachter 1991; Pallemaerts 1994). Earlier conferences and activities of regional or functional organizations possibly did as much to promote specific actions to protect the environment, but Stockholm began, or at least made explicit, the process of a global response to care for the Earth's ecosystems under a common framework. For example, scientists and diplomats introduced many of the concrete recommendations that came out of Stockholm at UNESCO's Biosphere Conference in 1968. However, the political dynamics of Stockholm were without precedent (Adams 1990:32–36). In the words of one analyst: "Stockholm was without doubt the landmark event in the growth of international environmentalism. It was the first occasion on which the political, social, and economic problems of the global environment were discussed at an intergovernmental forum with a view to actually taking corrective action" (McCormick 1989:88). This interaction of science, public concern, and international politics produced the first real clues as to how the international community as a whole would treat environmental concerns.

The Stockholm Conference was first and foremost an environmental conference. Held June 5–12, 1972, it brought together 113 states, 19 intergovernmental agencies and about 400 NGOs in the parallel Environment Forum. The only notable absences were members of the Soviet Bloc who boycotted the conference in protest over the exclusion of East Germany.[3] Significant outcomes of the conference included the 26 principles of the Declaration on the Human Environment, the 109 recommendations in the Action Plan for the Human Environment, and the creation of UNEP, formally established by the United Nations General Assembly in December 1972 (United Nations 1972a).

I am concerned mainly with the Declaration principles. These principles brought together the interests of the developed and developing world, thus highlighting the tension between environment and development. By forcing that conflict into the open, Stockholm marked a significant step in the development of the current norm-complex of liberal environmentalism. However, Stockholm did not work out the environment/development tension under a unifying set of norms. Rather, the final documents simply juxtaposed the interest in environmental protection by the North with the development concerns of the South. The Declaration and Action Plan introduced macroeconomic issues related to trade and development, but never clearly specified the content of development norms it could support. Before exploring these norms in detail, I trace the dynamics that led to the Stockholm outcome.

Bringing the Developing World In

The Stockholm Conference was prompted mainly by concerns in industrialized countries over transnational pollution from industry and its products. Most accounts gloss over its unusual origins in the person of Swedish soil scientist and television personality Svante Odén. An advisor to the Swedish government, Odén played an instrumental role in convincing politicians and the Swedish people that lakes and rivers in Sweden were becoming acidic partly as the result of sulfur from smokestacks in other countries. Persuaded of the need for international cooperation to limit acid precipitation, Sweden proposed the conference.[4]

The time was also ripe for such an international gathering. By then environmental movements had sprouted up in many Western industrial nations; hence UNCHE reflected increased public anxiety over the state of the environment and the supply of natural resources. Popular publications such as Rachel Carson's *Silent Spring* and the Club of Rome's *Limits to Growth* fuelled those concerns (Carson 1962; Meadows et al. 1972). The former documented the effects of chemical insecticides on birds and other animal species (including humans) while the latter utilized a newly developed MIT computer-generated simulation that modeled trends of rising population and declining resource stocks. This early attempt at analysis of complex systems of cause-effect relationships predicted an impending resource crisis within 100 years if trends continued. The increased sensitivity to environmental problems, combined with spectacular environmental disasters such as the 1967 *Torrey Canyon* oil spill off the coast of Cornwall in Great Britain, contributed to the perception that environmental problems were severe, on the rise, and in need of a global response.

Not surprisingly, then, when in 1968 the United Nations proposed a global conference on the environment, the concerns of industrial pollution and the perceived need to conserve natural and biological resources initially dominated the agenda. Governments believed that such problems of industrialization required international cooperation and regulation. The initial United Nations resolution in December 1968 that called for the conference reflected this thinking, noting "that the relationship between man and his environment is undergoing profound changes in the wake of modern scientific and technological developments." The resolution identified physical effects of pollution and their immediate causes, but tended to see such problems in isolation from socioeconomic structures. Environmental concerns covered only "the continuing and accelerated impairment of the quality of the human environment caused by such factors as air and water pollution, erosion and other forms of soil deterioration, waste, noise and secondary effects of biocides, which are accentuated by rapidly increasing population and accelerating urbanization" (UNGA Res. 2398 (XXIII) A/L.533 December 3, 1968 in YUN 1968: 477).

Developing countries expressed concern over this narrow environmental focus. They successfully used the United Nations multilateral setting to demand the inclusion of development issues. In particular, many developing states feared an emphasis on lifeboat ethics or no-growth philosophy implied in studies such as *Limits to Growth* (e.g., de Almeida 1972:37–56; *Founex Report* 1972:12–13, 27). Some states also voiced concerns that high-profile pollution and disasters would overshadow links between environment, culture, and economics. In particular, developing countries worried that trade barriers would be erected under the guise of environmental protection. For example, food exporters who relied on chemical pesticides worried they would lose markets in the developed world if tough regulations were imposed. If developing countries were to participate—which was crucial to the success of the conference—these concerns could not be ignored.

The change from 1968 to 1972 in United Nations Economic and Social Council (ECOSOC) and General Assembly (UNGA) resolutions demonstrated the progressive movement of developing world concerns to the mainstream, if not the center, of Stockholm's agenda. Initially, resolutions marginalized developing countries. They treated developing countries concerns as an afterthought, merely expressing a general conviction that attention to the human environment is "essential for economic and social development." Developing countries used resolutions in subsequent years to shift the position of development on the agenda. For example, a 1969 resolution for the first time placed the concerns of developing countries within the main purpose of the conference. UNGA resolution 2581 (XXIV) "affirms" that the primary purpose of the conference remained "to serve as a practical means to encourage, and to provide guidelines for, action by governments and international organizations designed to protect and improve the environment . . . bearing in mind the particular importance of enabling developing countries to forestall the occurrence of such problems" (YUN 1970:393).

In response to the report of the first session of the conference's Preparatory Committee (PrepCom) in New York (March 10–20, 1970), resolutions went further in an attempt to balance environmental problems of developed and developing countries. In particular, ECOSOC resolution 1536 (XLIX) stressed the need to take into account "such environmental problems as are particularly acute in developing countries and relevant to their needs." More significantly, in terms of articulating a particular set of norms, the same resolution, "*Earnestly hopes* [that the conference] . . . will promote, in particular, the aims of the Second United Nations Development Decade by contributing to sound economic and social development" (YUN 1970:451). However, the General Assembly, which had until then used ECOSOC language in its own resolution (2657 [XXV]) on the conference, simply took note of the ECOSOC resolution. A North/South split was apparent.

By 1971 the split was in the open. That year marked a significant shift in how the United Nations would treat global environmental concerns: they no longer could be discussed in isolation from development. The General Assembly resolutions for the first time directly linked the two concepts, stating that "development plans should be compatible with a sound ecology and that adequate environmental conditions can best be ensured by the promotion of development, both at the national and international level" (UNGA Res. 2849 (XXVI) in YUN 1971:311–312)). Other relevant language in the resolution reflected political and development goals of the Group of 77 developing nations (G-77). In response, the United States and Great Britain voted against the resolution and all other developed countries, East and West, abstained. However, it still easily reached the necessary two-thirds majority for passage.

Developing countries used the resolution to forcefully argue two points: first, that "pollution of world-wide impact is being caused primarily by some highly developed countries . . . therefore, the main responsibility for the financing of corrective measures falls upon those countries"; second, that most environmental problems in developing countries resulted from underdevelopment itself. In addition, among the provisions, developing countries stressed that states must respect sovereignty over resources and internal economic planning, that environmental provisions must not impede development, and that measures must avoid any adverse consequences for trade, technology transfer, or development assistance. In particular, the resolution asked for additional financial and technical resources "beyond the resources already contemplated in the International Development Strategy."[5] The resolution also listed a number of specific requests that later appeared in modified form in the Stockholm Declaration. The unmistakable underlying theme was that developing countries wanted assurances that environmental concerns would in no way impede their development goals.

Despite the impression of unanimity among developing countries in United Nations resolutions, not all developing countries found common cause on every issue. For example, the People's Republic of China (a latecomer to the conference process) took a strong stand against the United States' involvement in Vietnam while other developing countries took positions against China's and France's nuclear testing programs. However, on the core environmental and economic positions, the developing world acted largely as a bloc in an attempt to maximize its political power. It took advantage of Northern concerns that, in the future, the South could be the locale of the world's worst environmental problems. Not surprisingly, leadership in the South fell to countries such as India and Brazil (and later China) that traditionally filled that role. However, their importance to the global environment gave them added bargaining power. For example, India's large population and Brazil's sovereign control over huge rainforests gave those countries both power and an interest in the outcome of Stockholm.

The position of developing countries did not arise in isolation from more general development goals that had taken shape during the 1960s in fora such as the United Nations Conference on Trade and Development (UNCTAD) and G-77 meetings. These organizations had begun to push for various reforms in international institutions and to the international management of production, trade, and finance (with an emphasis on aid). The proposed reforms eventually coalesced in the early 1970s in demands for a New International Economic Order (NIEO). The demands included sovereignty over resources (and a general entrenchment of the norm of sovereignty and territorial integrity), increased aid flows from North to South, commodity price stability, increased participation and voting power in international economic institutions, and restructuring of trade to allow greater access to Northern markets and exceptions to the norm of reciprocity.[6]

Following in this context, two key meetings in the lead-up to Stockholm articulated an emerging developing country position on environmental problems: the November 1971 Second Ministerial Meeting of the G-77 in Lima, Peru; and the meeting of the Panel of Experts on Development and Environment in Founex, Switzerland on June 4–12, 1971. Significantly, Founex came three months before the third PrepCom for Stockholm where the intergovernmental working group presented the first draft of the Declaration on the Human Environment. Since Founex had a more direct relationship to UNCHE, I will discuss it in more detail following a brief discussion of the Lima meeting.

The G-77 ministers' meeting is more interesting for what it did not say about the environment than what it did. Despite the flurry of United Nations activity around the issue, the environment only merited one small item on a lengthy agenda dominated by trade and financial matters. Hence, the final report contained only a brief statement on "The impact of environmental policies on trade and development" (Group of 77 1981 2:210). Apart from a general acknowledgement that all humankind[7] should be concerned about the environment, the statement focused on the negative effects of environmental policies for developing countries. It listed four concerns: environmental policies should not adversely affect development; specifically, environmental policies should not adversely affect the flow or terms of financial assistance, set new conditions on international trade nor obstruct any efforts "towards the sustained economic development of developing countries"; environmental trade barriers should be removed; and environmental policies in developed countries should facilitate development in developing countries.

As can be seen, the language of sustainable development already existed in demands of developing countries, but it lacked the same connotations promoted in the Brundtland report. Thus one must use caution in interpreting such language. For example, the economic program pushed by developing countries in the early 1970s, while growth oriented, certainly was not consistent with liberal economic regimes (Krasner 1985). "Sustained economic development" meant simply that development must make economic growth its priority, regardless of how it would be achieved. As I will demonstrate later, the use of sustainable growth language has reemerged in line with neoliberalism in international economic institutional arrangements. Thus the manipulation of discourse plays an important role in justifying a series of changes to the norms invoked in the name of "sustainable development" more generally.

The Founex meeting marked a more thorough attempt than Lima to articulate concerns of developing countries. UNCHE secretary-general Maurice Strong convened the meeting of 27 experts in the fields of development and the environment in an attempt to repair the rift between the developed and developing world on the focus of Stockholm. In conjunction with four follow-up regional seminars in the developing world, Founex succeeded on at least three counts. First, it allowed respected experts somewhat sympathetic to environmental issues to express concerns in an environmental forum that placed the developing world front and center. Second, it cemented the linkage between environment and development issues, with the assertion that they could be combined to optimize sound economic and ecological systems, even if the relationship remained vague and ill-defined. Third, those experts became valuable political assets who helped convince developing world leaders to send delegations to Stockholm, and to attend themselves.

The four regional conferences, co-sponsored by the United Nations secretariat and convened by economic commissions in Addis Ababa, Bangkok, Mexico City, and Beirut, also generated developing country interest. Environmental scientists and administrators from developing countries dominated the conferences. Although continued skepticism about the Northern emphasis on pollution control prevailed, the added influence these meetings provided to developing-country experts allowed them to push their governments toward a moderate position in the preparations for Stockholm. They also provided momentum for the building of domestic environmental administrative capacity and concern (Engfeldt 1973:403). These meetings and Founex prevented Stockholm from being a political failure.

Substantively, the Founex report supported the two conclusions that appeared in the 1971 UNGA resolution's language above. First, it noted that, "To a large extent, the current concern with environmental issues has merged out of the problems experienced by the industrially advanced countries." Furthermore, these problems largely resulted from a high level of economic growth with its attending negative consequences for local and global environments. While developing countries wished to avoid "mistakes and distortions" that resulted in the most severe negative consequences of development, the report strongly argued that environmental problems in the South resulted largely from underdevelopment itself (the second conclusion of the 1971 resolution). "They [environmental problems] are predominately problems that reflect the poverty and very lack of development of their societies. They are problems, in other words, of both rural and urban poverty . . . [and] can be overcome by the process of development itself" (*Founex Report* 1972:10).

The report listed a second set of problems related to the development process that required attention from developing countries. These problems included unemployment, urban growth, population growth without corresponding economic growth, and the threat of deforestation. Founex also acknowledged the potential problems associated with large-scale irrigation, use of pesticides, and industrialization in general (1972:12). The remainder of the report spelled out specific environmental concerns and policy recommendations.

The report's significance, particularly for my purposes here, lies less in its recommendations or responses to particular environmental problems than in its influence on norm creation. Its substantive influence lies in three areas. First, the report demonstrated that developing countries were concerned about environmental problems, but were deeply suspicious of how the international community would deal with such problems if treated in isolation from development. Second, the report differentiated the environmental concerns of developing countries from those of developed countries. Whereas developed countries wished to control byproducts of industrialization, developing countries' primary environmental concerns were disease, poor water quality and sanitation, nutrition, and poor housing.

Finally, the report presented environmental problems in the context of international norms consistent with developing country concerns, in particular those expressed in the Strategy for the Second Development Decade. For example, it emphasized the sovereign control of developing countries over their economic development and their own resources. That norm would later be entrenched in Principle 21 of the Stockholm Declaration. Hence, where conflict existed between the goals of social, cultural and economic development, trade-offs "can only be made by the countries themselves in the light of their own situations and development strategies and cannot be determined by any rules established *a priori*." Furthermore, the report reiterated this concern under a section on environmental policy formation: "The formulation of environmental goals, as indeed the formulation of economic and social policies in general, falls entirely and exclusively within the sovereign competence of the developing countries" (1972:11, 12).

Despite its emphasis on development, Founex did not present a specific set of development norms. It fostered an expanded notion of development beyond economic growth that included other social and cultural goals. However, it established no clear definition of development nor did it specify the relationship between broader social goals and economic growth. The achievement of this inclusive notion of development seemed to be taken as a matter of faith. The report only discussed trade-offs in the broadest sense and maintained a cautious approach to any measures that might limit short-term growth. For example, the report highlighted the opportunity for developing countries to house polluting industries (such as petroleum, pulp and paper, and chemical industries) from the North, and presumed that the worst environmental costs of such industries could be avoided. "Such a development," the report stated, "opens up an opportunity for the developing countries to move into some of these industries if their natural resource endowments, including relatively less used environmental resources, create a comparative advantage in these fields."[8] Founex also noted, "to the extent that these objectives [environmental, social, and cultural] support or reinforce economic growth—and it can be shown that some of them do—their place would be more readily established" (1972:11). Economic development still clearly took priority.

Assessments of the Founex meeting differ on whether it truly achieved a synthesis of environment and development concerns. For example, Adams suggests the meeting primarily served to allay developing world fears about the economic effects of environmental policy. "In fact the Founex meeting did not break new conceptual ground. It simply repeated the statement of faith that development and environment could be combined in some way which would optimize ecological and economic systems" (1990:37).

Similarly, an independent assessment of implementation done ten years after Stockholm suggested that Founex successfully made the interests of developing countries known, but Stockholm as a whole did not produce the resources or commitment necessary to address those issues. It also downplayed the conceptual contribution of Founex. "Although the Founex report represented a useful start for the continuing debate on environment and development, at the time of the Stockholm Conference the issue was still largely perceived as a choice between environment or economic growth."[9] Consequently, developing countries mainly argued that they needed additional resources and assistance to enable them to take the environment into account. On this, Stockholm did not deliver (The Agesta Group AB Sweden 1982:3).

In sharp contrast, McCormick (1989:92–93) argues that Founex produced a consensus, forged by development economists, that the environment is a critical dimension of successful development. Founex also destroyed the idea that the two concepts were necessarily incompatible. McCormick argues that Founex convinced participants of the widespread nature of environmental concerns and that they should not be a barrier to development, but part of the process. Nonetheless, representatives at Founex, to quote Maurice Strong (1977:166), also "made it clear that they thought under-development and poverty constituted the most acute and immediate threat to the environment of their peoples."

Founex also demonstrated much about the complex interaction between personal diplomacy, political and economic interests, and science in the lead up to Stockholm. Founex would not have occurred without Maurice Strong's leadership. One analyst credited his "patient missionary work" with developing country governments and experts for avoiding a major North/South rift. Strong used meetings such as Founex and a scientific meeting in Canberra to air out the strongest aspects of the North/South rhetoric. And, in the PrepComs and in discussions with developing world governments, he constantly emphasized the compatibility between environment and development (McCormick 1989:95).

Strong's influence extended beyond his role as a good organizer, facilitator, and negotiator, for which he received similar accolades at the Earth Summit. First, Strong's appointment as secretary-general of UNCHE signified the politicization of the Stockholm conference. His appointment came relatively late in the preparatory process when he replaced Jean Moussard, a Swiss biologist originally chosen in 1969 as the Director of Studies responsible for the Conference proceedings. By mid-1970 it had become apparent to the United Nations leadership that Moussard, though successful in gathering scientific data, would not provide effective leadership needed to make the conference a success. UN Secretary-General U Thant, with the support of UN Under Secretary for Economic Affairs Philippe de Seynes, who had appointed Moussard, picked Strong, who had been head of the Canadian International Development Agency (CIDA) and a former President of Power Corporation in Canada, a large energy and resources holding company. According to Strong, his appointment came about after Swedish Ambassador to the UN Sverker Astrom contacted him "through a mutual friend" after the Swedish government "began to worry" that preparations had made very little progress by early 1970.[10] Astrom then recommended Strong to de Seynes.

Strong was particularly suited for the job because of his personal connections with developing country leaders through his work at CIDA, but also his earlier professional and voluntary activities. His proven commitment to development gave him credibility and respect in many countries in the South (Herter, Jr. and Binder 1993:12–13). For example, early in the preparations he met personally with Indian leader Indira Ghandi, using contacts close to her that he had developed in his work with CIDA, and convinced her to attend. Given her stature and the leadership of India in the developing world, the promise of her participation greatly enhanced the prestige of the conference and made a developing world boycott unlikely (Strong 2000:126–127; Herter, Jr. and Binder 1993:26). Strong also played an instrumental role in getting China to participate through a direct appeal to People's Republic of China (PRC) leader Zhou Enlai. Although the Chinese presence created additional political wrangling, it also increased the credibility of the conference and strengthened the developing world view.[11] Whereas Moussard saw the problem of the environment as a scientific one, Strong recognized the pragmatic requirements of multilateral negotiations. His personal style succeeded in brokering compromises among disparate political interests, building trust, and creating momentum for agreement.

Strong also directly influenced how the problem of the human environment would be characterized. Almost immediately upon his appointment in January 1971, Strong convened a meeting of five or six experts at MIT, including Donella Meadows and Jay Forrester of *Limits to Growth*. Carroll Wilson, a friend of Strong's and one of the leaders of the Club of Rome at MIT, set up the meeting. Peter Thacher, of the U.S. Mission to the United Nations, also attended. In a published interview, Strong said, "Basically, our objective was to entrench the issue of the control of the environment with the economic-development process, both in developing and industrialized countries" (Herter, Jr. and Binder 1993:21). That meeting produced the slogan that summarized the Stockholm mission: "to protect and enhance the environment for present and future generations" (Herter, Jr. and Binder 1993:21). The U.S. influence shows through in that this slogan essentially represented a conservation ethic already present in the U.S. National Environmental Policy Act (1969). Its first goal is to "fulfill the responsibilities of each generation as trustee of the environment for succeeding generations." With an emphasis on intergenerational equity, this slogan presaged only the conservation side of sustainable development, not the integration of environment and development.

However, Strong also convened a meeting of development experts that he called "the single most influential meeting in terms of my development of the agenda" (author's interview). This meeting in New York specifically aimed to bring development onto the Stockholm agenda. It provided a forum to hash out many of the issues that would be aired more formally in the Founex meeting, and many of the participants overlapped. Strong asked Barbara Ward (Lady Jackson), a well-known British developmental economist, to bring together a small group of prominent development experts. Most of the experts were economists from the developing world, many of whom continued to have a major influence on environment and development governance. The group included Gamani Corea of Ceylon (later secretary-general of UNCTAD), Mahbub ul Haq of Pakistan (later positions included a World Bank vice president, finance minister of Pakistan, and architect of the United Nations Development Programme human development reports), Abdlatif Y. Al-Hamad of Kuwait, Inrique Iglesias of Uruguay (who served as foreign minister, later headed the Brundtland Commission's advisory panel on energy, and then worked at the Interamerican Development Bank), and James Wolfensohn (who at the time of writing is president of the World Bank). Strong said his "whole thesis" when he agreed to run UNCHE was the need to integrate environment and development and this meeting helped to formulate how that could be done to reshape the Stockholm agenda.

The Declaration on the Human Environment

Of the outcomes of UNCHE, the Declaration best expresses the norm-complex that emerged and the compromises it embodied. The final draft declaration had changed in purpose and substance from its original conception in March 1970 as a largely educational and inspirational document of basic principles. By the third PrepCom in September 1971, the influence of Founex and increased public attention had combined to put pressure on the intergovernmental working group to produce a document that represented concrete action (Rowland 1973:87).

The first move in this direction came from an early Canadian draft. It proposed a legalistic document that listed substantive norms and principles that could be a basis for international law (reproduced in Rowland 1973:88). The principles included norms of sovereignty and state responsibility for pollution produced within one's own territory that caused damage in other states or in common areas beyond national jurisdiction. In addition, states whose pollution harmed neighbors would be obligated to compensate them and would be required to consult neighbors when such pollution was likely to result. Although the final draft declaration would retain these principles relatively unchanged in Principles 21 and 22,[12] the Canadian document, stated in terms of rights and obligations, said virtually nothing about the relationship between environment and development.

By the end of the conference, however, the environment/development compromise played a central role, while negotiators watered down the strict legal language of rights and obligations. Although some of those changes occurred before the intergovernmental working group handed the draft over to the fourth PrepCom in March 1972, many occurred afterward in the more politicized atmosphere of the conference itself.

The Chinese delegation played a major role in reopening the Declaration to amendments and discussion. Its motives ranged from a simple desire to be heard (the PRC had been left out of deliberations by the intergovernmental working group) to an attempt to use the Declaration for ideological purposes. In the end, the PRC did play a positive role in reintroducing many of the development issues that appeared in working papers, such as the Founex Report and Report by the Secretary-General on Development and Environment largely based on Founex.

The Chinese delegation presented a ten-point statement to the draft committee and also leaked it to the press through an NGO newspaper, ECO. The first point brought forward a view of environment and development that epitomized the uneasy meshing of concepts that characterized the current stage in international environmental norm creation. The statement on the "relationship between economic development and environment" read as follows:

> Economic development and social progress are necessary for the welfare of mankind and the further improvement of the environment. The developing countries want to build modern industry and agriculture to safeguard their national independence and assure their development. A distinction must be made between these countries and a few highly developed countries. The environmental policies of each nation must not impede development (Rowland 1973:92).

Other points included a statement that downplayed the then popular cataclysmic forecasts on population growth and called for moderate national responses, such as control of urban population and family planning; national sovereignty over resources; a proposal on pollution compensation (in line with the original Canadian proposal); and a proposal for technology transfer. Although subsequent discussions sometimes broke down into North/South rhetoric and acrimony (with the U.S. taking the hardest line against the Chinese proposals) and specific proposals caused splits within blocs as well, a consensus gradually emerged on many of the key issues. For example, Canadian and Chinese positions overlapped on many of the core legal principles and some developed states actively supported Chinese and African positions on development issues. Traditionally divisive issues such as colonialism, nuclear weapons, and the war in Indochina sometimes appeared ready to sabotage agreement, but in the end did not have a substantial impact on the general consensus achieved in the final declaration.

Negotiations over the final wording in many cases came down to incorporating developing country proposals, particularly China's, into the wording of the draft declaration. For example, language in paragraph four of the preamble came from a Chinese proposal that identified underdevelopment as the cause of most environmental problems in the developing world. Similarly, a Chinese proposal changed the emphasis in paragraph five on population from a position that "excessive population growth can defeat man's efforts to preserve the [E]arth's environment" to a position that identifies people as "the most precious" of things in the world and the source of social progress and wealth, while acknowledging "problems" that can accompany population growth.[13]

The final negotiations also moved the Declaration more toward distributive policies and away from a strict focus on conservation. For example, Principle 5 states that "The non-renewable resources of the [E]arth must be employed in such a way as to guard against the danger of their future exhaustion and to ensure that benefits from such employment are shared by all mankind." Early drafts concerned only conservation of resources until a Pakistani proposal at Stockholm added the second part, that would "ensure that benefits . . . are shared by all mankind." As Sohn points out, this language of distributive justice was consistent with language in other declarations, such as the 1970 UNGA Sea-bed Declaration. Similarly, Principle 10, newly proposed at Stockholm by nine African states, notes that "stability of prices and adequate earnings for primary commodities and raw material are essential to environmental management" for developing countries. These principles fit with the general thrust of Principle 2 as well, which implies a duty to preserve the Earth's resources for the benefit of all people.

The principles can be divided into three general categories: conservation; development; and state sovereignty and responsibility.[14] As the above discussion suggests, the themes sometimes overlap within various principles, reflecting compromises worked out during the course of negotiations.

Principles 1–7 primarily delineate facets of human activity that require attention for conservation and environmental protection. Principle 1 is a general statement about the responsibility to preserve the environment for "present and future generations" (although it also contains admonitions against apartheid, discrimination, and foreign domination). Principles 2–7 cover specific aspects of that responsibility, from preserving wildlife (4) and natural resources (2), both renewable (3) and nonrenewable (5), to pollution concerns on land, in the air (6) and in the seas (7). As mentioned, some of these principles were altered to stress a greater emphasis on distributional concerns, not simply conservation.

Principles 9 through 13 specifically address concerns of development in developing countries. Most significantly, Principle 9 calls for "accelerated development through the transfer of substantial quantities of financial and technological assistance" as the best response to environmental problems in the developing world. It also directly links environmental vulnerabilities to underdevelopment. Principle 10 asserts that stability of commodity prices is essential for developing countries to manage the environment effectively. Principle 11 admonishes states against the use of any measures to protect the environment that could adversely affect development or the ability to raise the standard of living in developing countries. Principle 12 calls for additional financial and technical assistance (above other development aid) for environmental protection in developing countries. Principle 13 places development as the primary concern in planning, but says it should be ". . . compatible with the need to protect and improve the human environment. . . ."

Principles 14–20 do not fit into the three categories above, but deserve brief mention to keep the remainder of the Declaration in context.

Principles 14–17 focus on national and regional planning. The significant aspect of these principles is the faith they conveyed in the ability of "rational planning" (14) to reconcile the needs of development and the need to protect the environment. Principle 15 calls for planning in human settlements. Principle 16, on population, is a weak statement that governments, based on their own priorities and without prejudice to human rights, should apply demographic policies (either to decrease or increase populations) as it suits environment and development goals.

Principles 18–20 focus on scientific research and public education. Principle 20 also includes a call for the transfer of information, experience, and technology to developing countries, without economic burden, to facilitate research and development.

Principles 21 and 22 contain statements on rights and duties of states. As previously mentioned, the original desire of some states to create strict rules of liability did not materialize in the final document, although the basis for the future development of such rules remained. (Nonetheless, little development of rules of liability in international environmental law has occurred since Stockholm, as will be shown later in the section on UNCED.) Environmental lawyers identify Principle 21 as the key norm for modern environmental law (Sands et al. 1994:7; Schrijver 1997). In conjunction with the United Nations charter and various General Assembly resolutions (e.g., 1803/62 of 14 Dec. 1962 on Permanent Sovereignty over Natural Resources) it cemented the notion that states have "a sovereign right to exploit their own resources pursuant to their own environmental policies." It advanced earlier United Nations resolutions because it also creates a responsibility on the part of states to ensure that their activities do not cause environmental damage beyond their own jurisdiction. Principle 22 originally meant to create liability from states that cause environmental harm beyond their borders and a duty by them to compensate the victims of pollution. However, the final version only requires states to "co-operate to develop further the international law regarding liability and compensation. . . ." As noted in endnote 12 to this chapter, another principle originally proposed that would have created an obligation for states to notify others of activities that might cause environmental damage did not make it to the final declaration.

From a strict legal standpoint, it should be noted that the norms embodied in Principles 21 and 22 did not originate with Stockholm solely, nor did the nonbinding Declaration create a consensus on their precise status in international law. For example, norms of state sovereignty over resources have roots both in widely accepted rules around sovereignty and territorial integrity as well as in various United Nations declarations and decisions of international tribunals that say that states have a responsibility not to cause damage to the environment of other states. This responsibility has been acknowledged at least as far back as the widely cited Trail Smelter case (1941), when an arbitration tribunal found Canada was responsible for damage in Washington State caused by fumes originating at a smelter in British Columbia.[15] The Stockholm Declaration itself is considered soft law, which in recent history often represents a first step for new areas of international law to be accepted by states as customary law. Regardless of its origins, much of the Stockholm Declaration, especially Principle 21, is now considered customary international law.

Principles 23 and 24 are not easily categorized. The former generally recognizes concerns of developing countries that each state can determine its own environmental standards based on its own values, but also emphasizes that standards "which are valid for the most advanced countries . . . may be inappropriate and of social cost for the developing countries." In this way it fits with Principle 21 and the general concerns of developing countries covered in other principles already mentioned. Principle 24 calls for international cooperation through bilateral and multilateral arrangements to protect the environment.

Finally, Principle 25 calls for the support of international environmental organizations, and Principle 26 calls on states to eliminate nuclear weapons.

The Norm-Complex

The preceding discussion indicates that a weak norm-complex of *environmental protection* resulted consistent with the view of Western environmentalists that development and environmental protection are different, often competing tasks, the latter being concerned with regulating "externalities" (Colby 1990:8). It included an uneasy mix of conservation, economic development, sovereignty, and state responsibility norms, but essentially highlighted the incompatibility of many development and environmental goals, not a synthesis. The Action Plan reinforced this view. Only eight of 109 recommendations address development and environment, and are stated primarily in the negative, that is, environmental policies should not harm development, trade, and so on (Adams 1990:39; United Nations 1972a,b). The Stockholm outcomes as a whole contain goals of conservation and environmental protection side by side with a vi-

sion of development consistent with G-77 formulations, but lack any specification of trade-offs or how to make linkages.

Since the significance of these outcomes is that they embody a nascent
form of the current norm-complex, the following list identifies the norms
most relevant to this evolving norm-complex, organized under the categories listed earlier. The list nonetheless indicates that while the germs of
the current norm-complex of liberal environmentalism are present, multiple pathways to very different futures could also be imagined.

State Sovereignty and Responsibility:

1. States have sovereignty over resources and environmental protection
   within their jurisdiction and are responsible for pollution they produce beyond their borders (Principles 21–23).

Political Economy of Environment and Development:

2. The sources of environmental problems differ in developed and developing countries and so should responses.
   a. In developing countries, accelerated economic and social development (which are not specified) are compatible with and necessary
   for environmental protection (Principles 11, 12, and 13).
   b. In developed countries, industrialization and technology require
   regulation to protect the environment.
3. Free trade must be balanced with commodity price stability (Principle 10).
4. Environmental protection requires substantial transfers of financial
   aid, technology, and scientific information to developing countries
   (Principles 9 and 20).
5. States should cooperate to conserve and enhance the global resource
   base for present and future generations (Principles 1–7 and 24).

Environmental Management:

6. Command-and-control methods of environmental protection are
   favored over market allocation. The integration of economics and environment is limited to "rational planning," which is left ambiguous in
   meaning.[16] This last norm seems to apply to national and international
   planning. However, the economic and social implications of planning are
   not clearly specified.

Since I am concerned with norms of governance, I have not detailed specific recommendations for environmental protection activities. The action plan is significant for the purposes here only in so far as its recommendations do indeed reinforce the normative framework of the Declaration. I do not mean to downplay the importance of the specific priorities of the action plan, which contains, among other things, recommendations on pollutants to monitor, facets of human settlements that require attention, and a framework to manage natural resources. Rather, my focus has been on the international community's overall attempt to govern such activities, not the targets of action or the effectiveness of environmental protection *per se.*

Despite UNCHE's mix of environment and development, international environmental law and practice following the conference primarily emphasized the environmental protection side of the norm-complex. Developing countries were slow to embrace the environmental protection norms promoted at Stockholm while developed countries focused attention mostly on pollution abatement and clean-up at home. A detailed independent study on the implementation of the Stockholm proposals ten years later found that "The expectations and objectives of the developed countries were largely achieved at and after the Stockholm Conference but, of course, to varying degrees" (The Agesta Group AB Sweden 1982:3). (Although, the report also lamented the decline in political will to address environmental problems in North and South alike by 1982.) Developed countries focused on two priorities: identification and control of pollutants of broad international significance and environmental aspects of natural resource management. In contrast, developing countries received "no significant" additional financial resources to help them deal directly with environmental problems. Despite the success of incorporating developing country concerns into the Declaration, "the issue was still largely perceived as a choice between environment or economic growth."[17]

The above list of norms will serve as the point of comparison when I enumerate the norm-complexes that emerged from the Brundtland Report and UNCED.

FROM STOCKHOLM TO SUSTAINABLE DEVELOPMENT

On the path from Stockholm to the Rio Earth Summit, "sustainable development" emerged as the dominant conceptual framework for international environmental governance. The set of norms produced at Stockholm lacked a unifying theme either to forge a consensus between North and South or to capture the imagination of world opinion. Sustainable development meant to change all that. In one concept, environmentalists, economists, planners, industrialists and governments of all political persuasions could find a unity of purpose, if not agree on how that might be accomplished. As one author put it:

> It is not surprising that such a concept has received widespread support from leaders of the North and South alike, environmental and Third World movements, international bureaucrats and enlightened managers of financial and economic institutions and structures in both capitalist and socialist countries. This is explained by the artful vagueness which the new paradigm of 'sustainable development' casts upon their respective responsibilities (Pallemaerts 1994:14).

Its vagueness, rather than condemning it to the trash heap of development concepts, made it the favored mantra of international environment and development communities.

This section focuses on the evolution of the language of sustainable development in international discourse, its sources, and its eventual delineation in the prominent World Commission on Environment and Development. Divergent paths from Stockholm, roughly corresponding to Northern and Southern positions, eventually converged around the "sustainable development" concept. Two developments in 1974 set the normative character of those paths: the development of guidelines for the Polluter Pays Principle (PPP) in the North and the Cocoyoc Declaration in the South. Whereas international cooperation in the North focused on methods of internalizing environmental costs with minimum disruption to markets, the South sought an overhaul of the international economic order, which it felt relied too heavily on the market to the detriment of the poor.

The North

The development by the Organization for Economic Cooperation and Development (OECD) in 1974 of guidelines to implement the PPP set the tone for the North (OECD 1974). The OECD originally developed the PPP two years earlier, not as a rule of liability, but as a means to avoid environmental regulations that might alter the operation of the market and particularly of free trade (OECD 1972, 1975). OECD recommendations in the early 1970s gave PPP a restricted meaning that said that pollution abatement by the private sector should not be subsidized by governments, which would create a burden on the wider community and could distort trade. Properly implemented, PPP would ensure that market prices more closely reflected the social costs of production. However, PPP also implied internalization of environmental costs—a meaning initially downplayed in its implementation, but which subsequently took hold and now dominates.[18] This meaning can be seen even in its earliest formulation, which states that polluters should bear the cost of pollution they cause and resources they use "to ensure that the environment is in an acceptable state" (OECD 1972).

The introduction of the PPP marked the start of a trend to incorporate environmental costs into production, markets, and accounting practices, rather than favor what economists label command-and-control regulation to combat environmental damage. Under PPP, public choice might determine the level of environmental protection sought, but, ideally, implementation of such standards would rely on the manipulation of market incentives, not strict end-of-pipe regulations. Admittedly, PPP in practice often took shape in the form of direct regulations based on standards, permits, and so on, which impose costs on meeting those standards to the polluter. Even in such cases, however, PPP relies on proper pricing so that market signals to consumers, for example, will reflect the full social and environmental cost of goods produced. Furthermore, the spirit of the principle implies the use of market-friendly instruments such as pollution charges and tradeable pollution permits, as evidenced by the trend in supporting such instruments in implementing the PPP in the 1980s and 1990s.

At the international level, the OECD intended the principle to "avoid distortions in international trade and other economic relations which might arise from differences in member countries' pollution control measures" (OECD 1972). Under this principle, subsidies, for example, would not accompany measures to implement the principle since they distort the market. The notion of "getting prices right," and the field of environmental or ecological economics that primarily concerns itself with this task, follow from this basic principle.[19]

Despite the intentions of its framers to limit the PPP to an economic principle, some developing countries have attempted to extend the principle to cover liability and equity concerns between North and South. For example, some interpret PPP to require developed countries, as the historic site of the majority of practices that damage the environment, to shoulder greater responsibility, and costs, for environmental preservation and management.[20] However, in policy and legal terms, PPP retains the more narrow meaning ascribed to it. Hence, other norms, such as that of "common but differentiated responsibility" supported at Rio, have been required to invoke the broader implications sometimes associated with PPP.

It should also be apparent that the underlying logic of PPP and similar mechanisms ensures support for economic growth. The developers of this principle believed that if environmental protection can be achieved with a minimum distortion of markets, economic efficiency and growth would be maintained, thus minimizing the need for trade-offs between growth and environment. As indicated above, the originators of PPP explicitly meant it to avoid distortions in international trade because that might limit growth.

The OECD's work in this area received a tremendous boost when, in 1984, Environment Director Jim MacNeill organized the "Environment and Economics" conference. The OECD economics establishment fully supported the conference, which helped to make it a major influence on governments and business in the direction of the OECD environment directorate's vision of environmental governance. The conference emphasized the desirability of strengthening the role of economic instruments and the reciprocal positive linkages between environmental protection policies and economic growth (OECD 1985). The pivotal role the conference and the OECD more generally played in legitimating these linkages, thus in influencing the future direction of environmental governance, is discussed further in chapter 5. In short, the conference helped to shift the way governments, business, and the economic establishment at the OECD thought about environmental issues and the best ways to address them. In particular, it cemented the view that economic growth and environmental protection could be compatible. MacNeill's later role as secretary-general of the Brundtland Commission ensured those ideas would also influence efforts at global governance for some time.

Major industrial states during this period, to varying degrees, also began to reformulate the importance and direction of environmental policy. Here, too, one finds that increased attention to environmental concerns led to increased efforts to find a fit between those policies and liberal economic norms. The general trajectory of European Community (EC) goals, for example, followed a similar pattern to that of OECD policy statements, although European policy lagged slightly in comparison.

From the start, Article 2 of the 1957 Treaty of Rome spelled out the fundamental objectives of the Community in economic terms. Whereas one would not expect this pre-Stockholm document to focus on environmental concerns, significantly the Single European Act of 1987, which contained a new chapter on a legal basis for community action on the environment, left Article 2 with its economic focus. While the Maastricht Treaty finally reformulated EC objectives, it nevertheless calls for the promotion of "a harmonious and balanced development of economic activities [and] sustainable and non-inflationary growth respecting the environment" (Title II: Provisions Amending the Treaty Establishing the European Economic Community With a View to Establishing the European Community, Article G paragraph B.2). Significantly, the language used in the Community debate on sustainability had been that of "sustainable development" at least following the Brundtland report. However, at the Rome Summit of December 1990, which considered the future work of the Intergovernmental Conference on Political Union, heads of government requested that the conference consider protection of the environment to ensure "sustainable growth." Despite some discussion of the change in terminology, the growth language stood during negotiations toward 1992 and no head of government tabled the issue at Maastricht (Verhoeve et al. 1992:14–15). The language of growth and open markets thus circumscribes the language of environmental concerns in EU documents.

This discussion should not imply that the EU lacks a serious concern for environmental protection within its borders or in its relationship to the developing world. In fact, since 1987 the EU has led the West in pushing forward the international environmental agenda. The above discussion merely points out that the EU's framing of environmental concerns rests on a primary concern with economic growth and that sustainable development is defined in such a way as to be compatible with growth and market forces.

The United States took an even stronger pro-market view, especially under the Reagan administration (McCormick 1989; Kraft and Vig 1984). A convincing case can be made that the shift in policy to deregulation, cost-benefit analysis, and heavier reliance on market incentives came from an ideological shift rather than an assessment of policy effectiveness. For example, Kraft and Vig (1984) traced administrative changes that consistently put political control of the environmental agenda above expert administration. A key turning point in U.S. policy came when the Reagan administration virtually ignored the report of a transition task force on the environment that it had set up. The report advocated moderate reforms that would ease some regulations, reexamine some laws, promote some economic incentives for environmental protection, but generally maintain the momentum for environmental protection. Instead, Reagan's environmental policy followed the much more radical position advocated by the conservative Heritage Foundation and Secretary of the Interior James Watt. He firmly believed in deregulation and that most resource problems could be solved by opening them up to the free market.

That view extended to U.S. foreign policy. In 1982, at a special session of UNEP to commemorate the tenth anniversary of the Stockholm Conference, Environmental Protection Agency Administrator Anne Gorsuch announced the unequivocal compatibility of growth, environmental protection, and markets:

> Individual ownership of property [and] free and well-developed markets in products and capital are powerful incentives for resource conservation. These institutions best promote the use of renewable resources and the development of substitutes for nonrenewable resources, ensuring continued resource availability and environmental quality (Gorsuch 1982).

The South

While Northern policymakers concerned themselves with methods to internalize environmental costs, the South in 1974 produced the Cocoyoc Declaration at a meeting in Mexico October 8–12. The Symposium on Patterns of Resource Use, Environment, and Development Strategies, billed as Founex II, brought together 33 delegates from eight developed and 14 developing countries (McCormick 1989:152). Maurice Strong, then executive director of UNEP, and Mostafa K. Tolba, who took over the post shortly thereafter,[21] also attended the conference, jointly sponsored by UNEP and UNCTAD. Founex II meant to further the work on environment and development started at Stockholm. Delegates discussed development strategies and international economic relations, analyzed environmental issues and the limits of natural resources in particular, and addressed the debate then occupying the United Nations where developing countries had just introduced the NIEO. The backdrop of the NIEO and the onset of the first oil shock a year earlier set the tone for the vision of environmental management that emerged.

A deep distrust of market mechanisms undergirded the Cocoyoc Declaration (UNEP 1981:109–119). It began with a stark summary of the lack of progress on poverty, hunger, illiteracy, disease, and homelessness, and the newer problems of resource degradation. It argued that the maldistribution of resources and overconsumption by the wealthy lies behind humanity's inability to meet the "inner limits" of satisfying fundamental human needs and the "outer limits" of the planet's resources. The solution, the report said, "cannot be left to the automatic operation of market mechanisms. The traditional market makes resources available to those who can buy them rather than those who need them, it stimulates artificial demands and builds waste into the production process, and even under-utilizes resources." The critique of the market extended to domestic systems of the time where the benefits of growth accrued to a small percentage of the wealthy while the poorest 20 percent grew poorer still.

The remainder of the Declaration set out the goals of development, which, it stated, should first provide for basic needs. Its recommendations fit with the vision of global economic management sought in the NIEO and the Charter of Economic Rights and Duties of States, both of which the Declaration explicitly endorsed. The Declaration's recommendations were as follows:

1. Governments, international organizations, and scientific communities should develop and institute policies that aim to satisfy the basic needs of the poorest and redistribute resources where possible. At the same time, they should ensure adequate conservation of resources and protection of the environment.
2. Within the framework of sovereignty over resources, governments and international institutions should promote the management of resources and the environment on a global scale.
3. Strong international regimes should be established for the exploitation of the global commons, and the use of the commons should be taxed for the benefit of the poorest strata of the poor countries.
4. Scientific and technological research and development should establish new priorities to respond to the goals of the report.
5. New development priorities should aim to curb overconsumption in the North and step up the production of essentials for the poor.

As can be seen, Cocoyoc placed the correctives to environmental problems squarely in the context of overall demands for a redistribution of resources. It had a bias toward global management of global resources and schemes for transfers from rich to poor to pay for the maintenance, equal access to, and use of global commons (for example, through taxation of the commons). Again, these proposals conflicted with liberal economic norms of free trade and market incentives and mechanisms for environmental protection and technology transfer.

Like the NIEO, however, the Cocoyoc Declaration had little lasting influence in terms of practical policy application. Nonetheless, it illustrated the state of development thinking at the time and how United Nations agencies concerned with development would view the environment agenda. As such, it marked the basis of environmental governance initially favored by the development community from the South.

UNEP's Role

Meanwhile, UNEP continued the Stockholm conference's work of reconciling environment and development. Indeed, by 1976, many delegates at UNEP's fourth Governing Council questioned the need to continue to defend the linkage of environment and development, which they felt had already gained wide acceptance (McCormick 1989:150). By 1980, the South appeared to verify this perception when it explicitly used the language of environmental sustainability in The Strategy for the Third United Nations Development Decade:

> It is essential to avoid environmental degradation and give future generations the benefit of a sound environment. There is a need to ensure an economic development process which is environmentally sustainable over the long run and which protects the ecological balance. Determined efforts must be made to prevent deforestation, erosion, soil degradation and desertification. International cooperation in environmental protection should be increased (UNGA 1980).

UNEP played a leadership role in developing this language. The secretariat, under Strong's leadership, worked to clarify the linkage between environment and development with a conceptual middle ground that emphasized economic growth, but of a "sustainable" kind. By the mid-1970s the language of sustainability (although not necessarily sustainable development *per se*) could be found in UNEP documents and speeches of its leaders. Strong thus could announce a solid support for economic growth, but of a new kind that considered the social aspects of development. "Economic and ecological factors must be brought into harmony in developing growth-patterns that are sustainable," he told the first International Environmental Management Seminar in 1975. "'Eco-growth' does not mean 'no-growth'; indeed it means better growth, sounder growth, and perhaps even more growth in qualitative terms."[22] He similarly called for a "'new-growth' society" in more forceful language two years later:

> Surely it must be clear that present growth-patterns and practices are self-destructive and cannot be sustained! Is no-growth then the only answer? Let me say with all the force I can muster that no-growth is *NOT* the answer. The real alternative to no-growth is new-growth—a new approach to growth, in both the more industrialized and the less-developed societies [emphasis in original].[23]

Language reminiscent of the G-77 Lima meeting's concern that environmental protection not interfere with "sustained economic development of developing countries" now had a positive environmental spin.

Although Strong labeled this vision "ecodevelopment," the "marriage" of ecology and economics, that term never really caught on with developing world governments. Nonetheless, United Nations agencies such as UNEP, United Nations Development Programme (UNDP) and the World Bank claimed to use the concept as a guide for incorporating environmental concerns into development planning. However, difficulties arose when translating the somewhat ideal language of ecodevelopment to the project level (Caldwell 1990:202–204).

Supporters of the concept point out its consistency with development thinking of the 1970s and its sensitivity to the complexity of ecosystems and how they respond to human interaction (Adams 1990:51–56). Themes such as local participation in projects, an emphasis on intermediate technologies, local self-sufficiency, and basic needs dominated academic and institutional writings in this vein. However, its failure to address broader debates about the global political economy, North-South conflict or questions of macroeconomic management likely accounts for its relative lack of success in capturing the interest of developing world politicians.

In general terms, ecodevelopment literature of the 1970s and 1980s overlapped with that on sustainable development. The main difference was an elimination of neo-Malthusian overtones contained in ecodevelopment's emphasis on small-scale development. UNEP had never really pushed that side of ecodevelopment, however, so the language and speeches of its leadership demonstrated the congruity of the concepts. Hence, ecodevelopment language merged with UNEP's later use of sustainable development terminology.

Shortly after Strong stepped down as UNEP's head, he stated that ecodevelopment "would be designed to assure that the precious natural resources . . . in the less-developed countries are exploited in ways that make the best possible use of their own skills and labor, and harmonize with their own culture and value systems to produce the resource-base on which sustained development depends."[24] An analysis of Tolba's speeches through the 1980s shows many of the same themes emphasized, but with sustainable development language fully substituted. Themes of his included repeated assertions of the interdependence of environment and development, the importance of poverty alleviation as a first priority, and an emphasis on a new qualitative evaluation of growth (Tolba 1987:97–107). Tolba also emphasized UNEP initiatives such as cost-benefit calculations and the general economic benefits of environmental protection.

In this way, UNEP took on the challenge of Cocoyoc but eschewed no-growth langauge. Tolba effectively used NIEO language while he avoided anti-market rhetoric that might alienate support from OECD countries. Sustainable development and sustainable growth became compatible concepts, even if pure growth in GDP no longer sufficed. Thus his submission to the Brundtland Commission, while it discussed many aspects of "sustainable development" consistent with the view of ecodevelopment and sustainable development above, emphasized economic growth as the basis of it all:

The first and most important premise [to put sustainable development into action] is the generally agreed perception that economic development and environmental quality are interdependent and, in the long term, mutually reinforcing. The rational management of the world's threatened natural resource base forestalls a loss in environmental quality and enhances sustainable economic growth (Tolba 1987:150).

All that was missing was a vision of governance to put such ideas into effect. The Brundtland Commission took on that task.

WORLD COMMISSION ON ENVIRONMENT AND DEVELOPMENT (WCED)

Two direct influences on WCED deserve mention to put the report in context: The World Conservation Strategy (WCS), often cited as the original source for the popular use of the term sustainable development; and a group of UN commissioned studies on development—Willy Brandt's *Programme for Survival* and *Common Crisis*—and security—Olaf Palme's *Common Security*.

UNEP commissioned the International Union for Conservation of Nature and Natural Resources (IUCN) to produce the 1980 World Conservation Strategy (IUCN 1980). The strategy intended to "stimulate a more focused approach to the management of living resources" and provide policy guidance for three groups: government policymakers and advisers, conservationists, and development practitioners, including aid agencies, industry, and trade unions. It received wide attention in those communities and proved somewhat effective in mobilizing national action on nature conservation. However, its lasting effect on norm creation at the international level was undercut by an inattention to political and economic factors that often lay behind stresses on living resources.

The final of three drafts, its authors admitted, was a compromise document. The IUCN prepared the document, but UNEP and the World Wildlife Fund (WWF), who financed the project, should be considered nearly equal partners as they played major roles in its preparation and influenced its themes and structure (IUCN 1980, ii). The United Nations Food and Agricultural Organization (FAO) and UNESCO also reviewed the final draft, which reflected wide consultations with interested parties from the conservation and development communities.[25]

The final strategy aimed to "help advance the achievement of sustainable development through the conservation of living resources" (IUCN 1980, iv). As such, it primarily focused on conservation of living resources, although some sections did mirror ecodevelopment thinking, that is, local development consistent with physical, biological and cultural resources, local participation, and so on.[26] It defined conservation as "the management of human use of the biosphere so that it may yield the greatest sustainable benefit to present generations while maintaining its potential to meet the needs and aspirations of future generations" (section 1.4). The definition comes close to Brundtland's for sustainable development, except the WCED replaced "management . . . of the biosphere" with "development." The strategy's definition of development, similarly, focused on the "modification of the biosphere and the application of human, financial, and living and non-living resources to satisfy human needs and improve the quality of human life" (section 1.3). The definitional linkages are clear—development, since it alters the biosphere, must take conservation into account to be sustainable. The solution, then, was to give conservation a higher priority.

The strategy contained three major objectives: (1) maintenance of essential ecological processes and life support systems such as soil, forests, agriculture, fisheries, and water; (2) genetic diversity; and (3) sustained utilization of species and ecosystems. It also contained a detailed set of priorities that addressed international concerns, but showed sensitivity to implementation at the local level.

Unfortunately for its supporters, WCS never overcame its lack of attention to the main concerns of developing country governments, nor did it take into account the essentially political nature of development. That problem, for example, meant an insensitivity to powerful interests in developing countries that favored rapid development and growth over environmental protection, or to pressures in the international and domestic political economies to exploit resources. Hence, many of the suggestions lacked context. As one analyst put it, "[WCS] seems to assume that 'people' can exist in some kind of vacuum, outside the influence of equality, class or the structures of power" (Adams 1990:51). On the bureaucratic level, it also ignored the planning process in many developing countries where central planning agencies, not environment ministries, controlled linkages to international development agencies. That, and the politics of those agencies, made it more likely aid would flow to conventional projects such as industry, energy, and agriculture rather than for the development of a conservation strategy (McCormick 1989:169).

Although WCS recognized North-South conflict in the international political economy, it did not successfully incorporate such concerns into the overall strategy. For example, while it explicitly endorsed a "new international economic order," in the same sentence it called for a new environmental ethic, stabilization of populations and "sustainable modes of development" (section 1.1). Similarly, its final chapters listed the demands of the NIEO without specifying why or how they fit with the conservation program in the WCS. It asserted compatibility of those values by definitional fiat: "Development and conservation operate in the same global context, and the underlying problems that must be overcome if either is to be successful are identical" (section 20.1). Then, after listing NIEO demands—a 0.7 percent official development assistance (ODA) target, better terms of trade, accelerated economic growth, and so on—it merely stated that, "Achievement of equitable, sustainable development requires implementation not only of the measures indicated above but also of the World Conservation Strategy." Finally, it urged that those conservation plans be included in the new International Development Strategy (section 20.5). Nowhere does WCS make the linkages between those aims explicit. Hence, its most lasting effect appears simply to have been the dissemination of the term sustainable development to governments and conservation advocates.

The Brundtland Commission meant to put sustainable development as it appeared in the WCS into a broader, development-oriented context. At the same time, it wished to further the multilateral and cooperative goals of the United Nations system. The Brandt and Palme commissions set the tone of Brundtland's broader objectives of multilateralism and interdependence. Gro Harlem Brundtland saw her task in *Our Common Future* explicitly as the third "call to political action" following on the two earlier projects on North-South economic relations and global security respectively. As such, she called the goal "to persuade nations of the need to return to multilateralism" as "perhaps our most urgent task" (WCED 1987:x).

The Cold War provided another important context. The Commission took on one of the few issues on which East and West could find common cause (Finger 1993:36–38). Already, organizations such as the International Institute for Applied Systems Analysis (IIASA) in Austria had provided a forum for such cooperation. Brundtland hoped to build on such efforts. The Commission stressed a "same boat" mentality. Images of a single, fragile Earth and interlocking ecosystems marked the introduction of *Our Common Future*. It followed that the planet's preservation required global environmental management and cooperation. Just as World War II produced the impetus for cooperation to build a postwar international economic system, "The challenge of finding sustainable development paths ought to provide the impetus—indeed the imperative—for a renewed search for multilateral solutions and a restructured international economic system of co-operation"(WCED:x).

The WCED Report

The significance of *Our Common Future* is threefold. First its high-profile origins as a UN General Assembly mandated project, unlike the WCS, mobilized sufficient public and political interest to elevate international concern on the environment. Until then, the priority accorded to environmental issues had largely declined in the wake of recession and debt since Stockholm. To generate interest and participation, the Commission's work included public hearings of senior government officials, scientists, other experts, industry, NGOs, and interested members of the public in all parts of the world (WCED 1987:359–361). The hearings generated more than 500 submissions, constituting 10,000 pages of material. In addition, WCED appointed expert advisers to assist the secretariat in analyzing key issue areas, set up advisory panels on energy, industry, and food security, and commissioned a series of reports from experts and research institutions. These efforts combined to give the report credibility and a high global profile.

Second, WCED cemented the linkage between environment and development that until then had been confined largely to communities directly involved in international efforts to promote such linkages. It also ensured that the relationship between environment and development would be framed in the language of "sustainable development." Third, it attempted to define the set of principles and norms that should underlie international efforts to achieve sustainable development. The first effect is self-explanatory, and the second two will be dealt with in turn.

Sustainable Development

The Brundtland Commission originated in a 1981 UNEP proposal to prepare an environmental perspective to the year 2000 and beyond (YUN 1982:1000). The following year UNEP recommended that a commission of eminent persons should help develop the perspective and mobilize public opinion. Finally, in 1983, after various consultations, the UNGA approved the establishment of a commission in resolution 38/161 without a vote. Its primary mandate was, "To propose long-term environmental strategies for achieving sustainable development to the year 2000 and beyond." Secondarily, its recommendations were to encourage cooperation between countries at different stages of development and to reach mutual objectives which "take account of the interrelationships between people, resources, environment and development."[27] Although WCED rephrased its mandate somewhat, the thrust remained to identify problems of environment and development and to formulate realistic proposals to address them.

The Commission, chaired by Norway's Gro Harlem Brundtland, worked in parallel to the UNEP Council's preparation of the Environment 2000 report.[28] The parallel process resulted not from a division of responsibilities, but from a turf war fought by the UNEP leadership who wanted control of the process.[29] That resulted in the marginalization of the UNEP report, whereas, by 1984, the work of the now-named World Commission on Environment and Development gained a high profile. Both reports were presented to the UNGA in 1987 and came to similar conclusions. However, WCED placed a greater emphasis on the growth side of sustainable development whereas Environment 2000 more often used terms such as "environmentally sound development," "effective environmental management," and "sustained environmental improvements" in the statement of its goals. The former view dominated future discussions, at least within the United Nations system, owing to the greater publicity and legitimacy granted to the WCED.

Our Common Future emphatically put environmental concerns in the context of an overall strategy of development. As Brundtland stated in the foreword:

> When the terms of reference of our Commission were originally being discussed in 1982, there were those who wanted its considerations to be limited to 'environmental issues' only. This would have been a grave mistake. The environment does not exist as a sphere separate from human actions, ambitions, and needs. . . .
>
> . . . the 'environment' is where we all live; and 'development' is what we all do in attempting to improve our lot within that abode. The two are inseparable (WCED 1987:xiii).

Chapter 1 of WCED further placed this philosophical position squarely in the context of the international political economy, stating that: "It is therefore futile to attempt to deal with environmental problems without a broader perspective that encompasses the factors underlying world poverty and international inequality" (WCED 1987:3).

Sustainable development was the cornerstone of WCED. Although defined variously, the most quoted definition reads as follows:

> Sustainable development is development that meets the needs of the present without compromising the ability of future generations to meet their own needs. It contains within it two key concepts:
>
> [1] the concept of 'need', in particular the essential needs of the world's poor, to which overriding priority should be given; and
>
> [2] the idea of limitations imposed by the state of technology and social organization on the environment's ability to meet present and future goals (WCED 1987:43).

Notwithstanding subsequent debates about the concept, WCED provided a relatively specific interpretation. First, needs refer to basic needs as defined by contemporary development discourse. Second, environmental limits are to be socially and technologically defined. Thus WCED framed the environment problematique in clearly cornucopian terms, a departure from the more eco-centric and conservation minded WCS.[30] In other words, according to WCED, decisions about limits must be made in the context of socioeconomic goals and what technology allows. The two documents came to similar conclusions on what environmental problems needed attention, but the rationale for concern differed significantly.

That difference was most clearly expressed in Brundtland's emphasis on growth. From its first page, WCED countered the limits to growth reasoning that pitted the developing world against conservationists at Stockholm:

> ... *Our Common Future*, is not a prediction of ever increasing environmental decay, poverty, and hardship in an ever more polluted world among ever decreasing resources. We see instead the possibility for a new era of economic growth, one that must be based on policies that sustain and expand the environmental resource base (WCED 1987:1).

The report made reviving growth the top strategic priority, in a sharp departure from earlier statements of global environmental policy. Specifically, WCED called for a minimum three percent annual increase in per capita income (which equals a five or six percent of GDP growth per annum) in developing countries and policies to redistribute income to alleviate absolute poverty. That rate would eliminate poverty in 25 years according to the report and would require "accelerated global growth" (WCED 1987:50–51, 89).

The report also recommended a better quality of growth: less material- and energy-intensive. Thus it reiterated the long-standing goals of UNEP noted earlier, while it played up those goals' foundation in economic growth. That foundation, it argued, led to the following other goals (after "reviving growth" and "changing the quality of growth") (WCED 1987:49):

- meeting essential needs for jobs, food, energy, water, and sanitation;
- ensuring a sustainable level of population;
- conserving and enhancing the resource base;
- reorienting technology and managing risk; and
- merging environment and economics in decisionmaking.

These goals undergirded detailed recommendations on reforms and priorities for incorporating sustainable development in the areas of food security, energy policy, urban development, living and nonliving resource conservation, population control and industry.

Two other goals were added in the Tokyo Declaration, made by the Commission in its final meeting on Feb. 27, 1987: to reform international economic relations, and to strengthen international cooperation. The former meant to prescribe the conditions for long-term growth. Specifically, the Declaration called for more equitable trade, capital and technology flows better synchronized with environmental imperatives, and fundamental improvements in market access, technology transfer, and international finance to help developing countries diversify their economic and trade bases and build self-reliance (WCED 1987:365). International cooperation applied to environmental research and monitoring and a general call toward multilateralism.

As these last goals indicate, the Brundtland report paid much more attention to international economic and institutional factors than did the WCS. It explicitly addressed the interactive linkages between poverty, environmental degradation, and macroeconomic relations. For example, it discussed the dependence of many African countries on commodity exports sensitive to declining prices. It also pointed to Latin America where debt crises and subsequent austerity programs had increased poverty and hurt distributional programs. Those governments faced pressure to make repayment a priority, thus they encouraged exports to generate foreign currency and pushed other development goals off or lower on government agendas. Such policies, WCED argued, are neither ecologically nor politically sustainable: "To require relatively poor countries to simultaneously curb their living standards, accept growing poverty, and export growing amounts of scarce resources to maintain external creditworthiness reflects priorities few democratically elected governments are likely to be able to tolerate for long." Furthermore, WCED argued that economic policies of some major industrial countries had depressed and destabilized the international economy, which aggravated these pressures on developing countries (WCED 1987:75).

The Commission favorably noted NIEO attempts to make economic arrangements more equitable, and to improve financial flows, trade, transnational investment, and technology transfer. Then, like WCS, it called for this program to consider ecological dimensions. However,

In the short run, for most developing countries except the largest[,] a new era of economic growth hinges on effective and co-ordinated economic management among major industrial countries—designed to facilitate expansion, to reduce real interest rates and to halt the slide to protectionism. In the longer term, more changes are also required to make consumption and production patterns sustainable in a context of higher global growth (WCED 1987:75).

Some modification of the international economic order would be necessary to achieve this synthesis of environmental concern and development. However, the basis of that order, WCED argued, should remain proper management by the major industrial powers. Hence, the goals of the international order should remain broadly liberal: interdependence, modernization, and free trade to promote economic growth.

The Norm-Complex

The Brundtland Commission promoted a governing norm-complex that encouraged a managed—or what might be loosely termed Keynesian—liberalism in the international economic order, infusing traditional forms of Keynesian intervention with an environmental bent. At the same time, explicitly environmental goals were to be incorporated into domestic development policies and in international institutions such as UNCTAD, the World Bank, and GATT to ensure that the economic order encouraged environmental concerns to be considered in decisionmaking and to prevent a growth-at-all-costs mentality.

The support of international Keynesian liberalism and interdependence remained largely unchanged from the Brandt reports. In this view, a sound global economy rests on free trade as the main engine of economic growth. However, selective interventions are accepted to propel developing countries into a more equitable position where they can better benefit from liberal economic institutions, or at least be cushioned from the impact of unfettered trade. Thus, on the one hand, WCED proposed a reduction in trade restrictions in the North, a reduction in real interest rates to ease debt payments, and an expansion of trade agreements to promote global economic growth. On the other hand, it supported increased financial flows in the form of aid from international development banks and other governmental sources, improved compensatory financing for commodity pricing to even out economic shocks, a strengthened bargaining position for developing countries *vis-à-vis* multinational corporations, and improved technology transfer arrangements, made possible by easing proprietary rights and encouraging joint ventures and cooperative research programs (WCED 1987:67–91).

It also gave the managed interventions it supported an environmental bent by, for example, proposing that increased aid should go toward projects that encourage sustainable development. The likelihood that such projects would involve higher local costs, a higher ratio of recurrent to capital costs, and a greater use of local technology and expertise, were not to deter lending in that direction. Such efforts might include "reforestation and fuelwood development, watershed protection, soil conservation, agroforestry, rehabilitation of irrigation projects, small-scale agriculture, low-cost sanitation measures, and the conversion of crops into fuel" (WCED 1987:77–78). WCED especially targeted the World Bank and IMF for reform since their lending conditions act as benchmarks for other governments and private lenders. Both their internal procedures and selection criteria ought to change, WCED argued, to reflect environmental and social costs and goals. Similarly, domestic policies in the North and South should be reoriented to resource conservation and enhancement.

A comparison with the norms of Stockholm will demonstrate the greater degree of synthesis achieved by Brundtland. However, it did not mark a radical departure. Indeed, it carried forward many of the same compromises on basic norms such as sovereignty over resources. Its difference lies in how it framed the norms of international environmental governance—differences that nonetheless are not insignificant as the new framing opened up avenues for substantial change in the legitimacy of environmental goals and the shape of policies and practices. Two changes stand out. First, for WCED, the synthesis of environmental and developmental goals suggested that governance of both rests on a common normative foundation, with economic growth at the center. Stockholm merely placed the two sets of interests side by side. Second, WCED explicitly spelled out the Keynesian-style compromise that ought to create obligations on the North for sound management and assistance, and responsibility on the South for reform. Below, the norm-complex promoted by WCED is presented with changes from Stockholm highlighted.

State Sovereignty and Responsibility:

1. Unchanged from Stockholm. Although a parallel legal process proposed new norms, they were never incorporated into the report nor were these proposals by a group of environmental experts from the North and South given serious consideration by the UNGA. When legal issues finally moved back onto the agenda in the PrepComs

for the 1992 Earth Summit, this set of legal principles did not form the basis of negotiations (Pallemaerts 1994:4; 1996:627–629).

Political Economy of Environment and Development:

2. The norm of differential obligations is downplayed. Instead, all states have a *common responsibility* to ensure a cleaner environment. Two imperatives apply equally to North and South:
   a. Revive global growth.
   b. Owing to environmental interdependence, require interdependent and shared responses to environment and development problems
      Poverty remains recognized as a source of environmental degradation in the South, and the North is seen to have an obligation to help alleviate it (through aid, and so on). However, a common program of freer and fairer trade to promote global growth combined with responsible regulation at the national level is called for in all countries. For example, developing countries have a responsibility to incorporate pollution costs into prices of pollution-intensive goods. Even the formerly taboo subject of sustainable population resurfaced in the report, albeit still in a weak form.

3. The international Keynesian-style compromise of balancing free trade with commodity price stability remains, although in more explicit terms. Free trade, and liberal economic policies generally, as the engine of growth, lies at the heart of the norm-complex. Managed interventions promote equity.

4. The argument that environmental protection requires substantial transfers of aid and technology for developing countries remains, followed by detailed proposals. Automatic financing, such as a tax on the use of global commons, is proposed in the spirit of the Brandt commission and the Cocoyoc Declaration (UNEP 1981; The Brandt Commission 1983:98–100). However, WCED made clear that political constraints made the implementation of such proposals unlikely in the near term.

5. The norm of cooperation to conserve and enhance the global resource base for present and future generations remains, with global growth a prerequisite.

Environmental Management

6. Encourage a mix of command-and-control regulation and economic/market-based incentives for environmental management. For ex-

ample, Brundtland's section on industry contained a discussion of economic instruments and recommended implementation of the Polluter Pays Principle (WCED 1987:219–232). Technological differences between North and South are to be considered (although WCED contained no mechanism to prevent PPP from penalizing industries from the developing world that may rely more heavily on subsidies for pollution prevention). However, economic instruments should be considered in the context of an overall strategy that also emphasizes standard setting, environmental assessment and government regulation. In addition, environmental audits should be required for transnational corporations that operate in developing countries.

Our Common Future legitimated the trend toward liberal environmentalism when it called for the integration of environment and economics in decisionmaking. However, the mix of management instruments and emphasis on various interventions in international markets left open a number of possibilities of how the ideas in Brundtland might eventually be institutionalized. Whereas WCED might call the norm-complex it supported "sustainable development," a better description is "managed sustainable growth." This will be contrasted with the liberal environmentalism of UNCED.

Chapter 3
ENVIRONMENT, DEVELOPMENT, AND
LIBERAL ENVIRONMENTALISM

THIS CHAPTER CONTINUES the story of the evolution of global environmental norms following the rise of "sustainable development" as a way to frame responses to global environmental problems. The designed ambiguity of that term meant it could hardly be considered a norm, according to my definition, in its own right. Yet, the attempts by states, international organizations, and nongovernmental actors to put sustainable development into action, programs, or treaty commitments started to coalesce around an identifiable set of norms by the early 1990s. That process culminated in the Earth Summit conference, the main focus of this chapter. Whereas many observers wonder if the Earth Summit succeeded in actually shifting international and local activities in a more sustainable or environmentally-friendly direction, it did succeed in defining how global environmental problems should be understood. Moreover, it delineated a range of appropriate behaviors and policy practices for dealing with global environmental problems that then set the pattern for action on specific environmental problems, and rights and responsibilities of the actors involved.

In the pages that follow, I show what happened following publication of the Brundtland Commission report that led to the Earth Summit's institutionalization of liberal environmentalism. I also describe the political dynamics of the conference process itself. The story of liberal environmentalism of course did not end with the signing of agreements at Rio. Its importance in the long run depends on its actual effects on global environmental practice. The chapter thus concludes with a preliminary discussion of how the institutionalization of liberal environmentalism shaped various efforts to either address global environmental problems or influence behavior within states that might be perceived as having transnational or global consequences. The argument is not that all environmental agreements or national policies now reflect liberal environmentalism, but that it has become the legitimate way to address global environmental problems, and a mainstay of how international organizations and states understand their role in promoting action at both the international and domestic levels. It remains to be seen both whether it will be effective, and how durable it will be in the face of a variety of potential challenges.

FROM BRUNDTLAND TO RIO

By the late 1980s it became apparent that although Brundtland's norms might appeal to the enlightened best intentions of the commissioners and some governments, WCED had been outpaced by the realities of the international political economy. In the North, fears of a surge in trade protectionism coincided with the rise of monetarist and neoclassical economic thinking, while many Southern states faced the rigors of structural adjustment. This combination made the success of the more radical redistributional proposals of WCED unlikely.

Instead, the IMF and World Bank programs to combat developing country debt began to reflect an emergent economic policy convergence, dubbed by John Williamson the "Washington Consensus" (Williamson 1990, 1993). Williamson meant by this a consensus among the "economically influential bits of Washington, meaning the US government and the international financial institutions," on the best course for economic policy.[1] Paul Krugman suggests the members of the consensus might also include "think tanks, politically sophisticated investment bankers, and world finance ministers, all those who meet each other in Washington and collectively define the conventional wisdom of the moment" (Krugman 1995:29). Krugman neatly summed up this ideological shift as "liberal trade and sound money." As originally articulated by Williamson, the Washington consensus included a call for states to liberalize trade and investment, privatize state enterprises and more generally encourage the retreat of the state from the economy, balance the budget, and peg the exchange rate.[2]

The success of the export-oriented economies in East Asia gave ammunition to promoters of structural reform, as did investor returns in those economies, even if states in the region did not uniformly embrace all elements of the "consensus." Moreover, when formerly communist states embraced the new liberal market orthodoxy, it seemed the end of the Cold War had ushered in near universal agreement on global economic norms.

Chapter 5 goes over the significance of this normative shift in the social structure of the international political economy. Whatever the causes of this shift, one effect was that global negotiations no longer inevitably stalemated over intractable differences in economic norms. As will be discussed further below, the consensus did not mean the South gave up calls for greater global equity or redistribution. Nor did government reforms in a neoliberal direction occur without significant societal opposition in some cases. Reforms also often occurred under significant financial duress and pressures from international financial institutions that reflected power differentials exacerbated by external debt and liberalization of financial markets globally. Regardless, whether in response to these external pressures, perceived shifts in the global political economy, and policy failures of earlier development strategies such as import-substitution, or a true conversion of financial and government elites or powerful commercial interests in many developing countries and economies in transition—or more likely a combination of the two—the public position and negotiating stance of developing countries no longer reflected a resistance to global liberalism (Biersteker 1992; Rodrik 1994; Busumtwi-Sam 1995).

One important indication of this shift became evident in the "Cartagena Commitment" agreed to at UNCTAD VIII, just four months prior to UNCED (United Nations 1993). The agreement among UNCTAD governments cautiously endorses market reforms along the lines of the Washington Consensus both in terms of domestic macroeconomic policies and in trade and foreign investment policies. It notes, for example, that although sometimes necessitating difficult policy choices, "it is becoming increasingly evident that a number of countries implementing those reforms are beginning to see the benefits" (United Nations 1993:9). In addition, UNCTAD, as the agreement and subsequent research and program development indicate, has shifted to a much more serious engagement with environmental issues under the rubric of incorporating "sustainable development" into the mainstream of its activities (Arda 1996). This shift in the international institution most associated with the New International Economic Order is perhaps the most remarkable outward indication that a profound shift had occurred in how developing countries viewed themselves and their place in, and understanding of, the international political economy.

The Earth Summit came in the midst of the apparent "triumph" of the Washington Consensus. By the mid-1990s, the successful completion of the Uruguay round of the General Agreement on Tariffs and Trade (GATT), financial deregulation, and increased efforts to liberalize regional trade all indicated that the North's response to protectionism fit with the emergent consensus, at least in foreign economic policy. Whether by will or by submission, these norms of global economic governance gained acceptance in North and South alike (Biersteker 1992).

The Brundtland Commission and Environment 2000 process paved the way for a coinciding transition in international environmental governance. The reports turned the corner on environmental thinking that had put it in direct opposition to classical economic views of growth and development. A UNEP official summed up the Environment 2000 report this way (and the same could be said for WCED):

> The [Environment 2000] Perspective also wants to remove fragmentation in thinking and action on human affairs, and it wants deliberate reconciliation of social, economic, and environmental aspects of human well-being in all countries. It wants the economic mechanisms of prices, charges, taxes, subsidies, allowances, permits, and rights to supplement regulatory frameworks, to bring about compatibility between environmental and economic objectives of development and private decisions, and systematic implementation of social development policies that contribute to environmental protection and improvement (Dabholkar 1989:53).

Not only was the compatibility of growth and environmental protection cemented in international discourse, but economic instruments and market-based solutions were already perceived to be the mechanisms best able to achieve this synthesis.

By pronouncing this compatibility of growth and environmental conservation and protection, Brundtland acted as a catalyst for a series of initiatives and research projects by those who wanted to develop the means to link what they saw as sound economic thinking with environmental protection. *The Economist* picked up this thread immediately in an endorsement of Brundtland's potential to, in its view, realistically ameliorate environmental degradation, "... if the eco-lobby could digest one of the study's least-trumpeted implications—namely that in most of the world economic growth and environmental protection go happily hand in hand" (*The Economist* 1987). The prescription was obvious in *The Economist*'s eyes—privatize the commons, create efficient markets for resources, and free capital markets for investment and lending:

> That is where the rich countries and their lending agencies should come in. With the right incentives in place, they need not worry about the clash between growth and the environment. The World Bank and the IMF will be doing the environment a favor when they insist on freer markets in exchange for their money (*The Economist* 1987:16).

Sustainable development, in this view, found no contradiction with the neoclassical turn in international economic governance.

The World Bank quickly picked up on this theme with a series of reforms begun in 1987. The reforms provide a logical starting point for an analysis of the direction of international environmental governance following WCED. The Bank's privileged position as a funnel for wealthy states' development funds, and especially concessional lending, along with the IMF, meant regional and commercial banks often mirrored its development policies. Thus the Bank is generally recognized as the premiere international development institution. Furthermore, the so-called greening of Bank policies not only produced a change in its lending practices, but also presaged a major foray into global environmental management that successfully culminated in its senior partner role in the new Global Environmental Facility or GEF (World Bank, UNDP, and UNEP 1992). Established in 1991 in partnership with the UNDP and UNEP, the GEF is now the main multilateral source of funding for major global environmental agreements and for disbursing monies attached to initiatives agreed to at the 1992 Earth Summit.[3]

An examination of the Bank requires sensitivity to its two-way relationship with environmental governance. On the one hand, Bank officials have made concerted efforts to reform the institution to make it more sensitive to the environmental consequences of its loans.[4] On the other hand, the Bank has played an active role in framing the norms of environmental governance.

The Bank generated the most publicity with its internal reforms, probably because of the intensity, volume, and sophistication of the criticism in the 1980s that presaged change. Those criticisms came not only from prominent NGOs and grassroots organizations in developing countries, but also from public pressure in industrial shareholder nations on which the Bank depended for its capital. In particular, the U.S. Congress held more than twenty formal hearings on the Bank's environmental policy. The U.S. concern culminated in its refusal to support a Brazilian power-sector loan in 1986, the first time it had voted on environmental grounds (Goodland 1992:11; Rich 1994:136–138). Bank staff had also started to notice that serious environmental degradation had begun to constrain development and undermine Bank projects, and evidence mounted that loans in many cases had themselves caused major environmental disasters. Although it hired its first (and at the time only) environmental adviser in 1970, significant reform waited until 1987 when the then new president, Barber Conable, made a well-publicized speech on May 5 at the World Resource Institute in Washington. He announced a major reorganization of the Bank, including the augmentation of the one weak environmental division into four regional divisions and one central department. That meant a sixteenfold increase in environmental staff, to about 100 people. Then, in 1989, the Bank adopted an environmental assessment umbrella policy (Goodland 1992:10–12; Rich 1990, 1994:145–181). Reforms continued, including the August 1994 initiation of a new inspection panel, which allows affected parties to launch reviews of whether the Bank follows its own policies, procedures, and loan conditions (Hunter and Udall 1994). Although environmentalists and Bank officials may disagree on the effectiveness of such reforms, the monitoring and assessment of environmental consequences has clearly increased and some movement has been made to include environmental considerations into assessments of project viability and impact.

Nonetheless, the second aspect of World Bank activities—the promotion and implementation of environmental norms—is where the Bank has had a more general impact on global governance. The most accessible and widely distributed statement of that policy can be found in the influential 1992 World Development Report, on the theme of environment and development (World Bank 1992b). Like the Brundtland Commission, the World Development Report argued that economic growth is the necessary condition for achieving other ends, including environmental protection and poverty reduction. The report projected a 3.5 times increase in world output between 1990 and 2030, and then argued that economic growth could be achieved without environmental deterioration, provided proper policies are in place. Proper policies, the report argued, are those consistent with goals prescribed in previous development reports, namely "market-friendly" policies for development (World Bank 1992b:9–10).

Hence the Bank's four-pronged program for "sustained development" (its preferred term because it narrows Brundtland's definition to "rising and sustainable levels of welfare")[5] began with two policies specifically aimed at market liberalization: first, "Removing subsides that encourage excessive use of fossil fuels, irrigation water, and pesticides and excessive logging"; and, second, "Clarifying rights to manage and own land, forests, and fisheries." These two policies essentially supported the Polluter Pays Principle in that they attempted to internalize environmental costs by eliminating subsidies and clarifying property rights.

The final two planks focused on establishing social conditions conducive to such reforms. The third plank promoted an acceleration of the provision of basic needs such as drinking water, sanitation, education (especially for girls), family planning, and agricultural extension, credit, and research. Finally, the Bank supported greater participation in development decision making at the community level. It should be noted that the Bank argued that even the provision of basic services could be best achieved by assigning property rights and other market reforms, which are presumed to limit pollution better than either common access or ownership regimes, or command-and-control regulations. "Market-based instruments are best in principle and often in practice," the report argued, to change environmentally damaging behavior (World Bank 1992b:2–3, 10–14).

Bank insiders echoed this general interpretation. For example, former director of the Environment and an author of the report, Kenneth Piddington, called environmental economics and proper valuation "the decisive element in the Bank's overall approach" (Piddington 1992: 222). Similarly, Mohamed T. El-Ashry, another former Environment Director and also Chairman of GEF, traced environmental degradation largely to inadequate property rights, subsidies for scarce resources such as water, polluting products such as pesticides, and other causes laid out above. On a macro level, he also called for the liberalization of trade and investment (El-Ashry 1993). Post-1992 World Bank environment reports continued to demonstrate an emphasis on the same liberal economic norms (e.g., World Bank 1994).

While WCED did not cause these changes in the Bank, it did play an important normative or enabling role. *Our Common Future* legitimated a form of international governance consistent with the Bank's general development philosophy—an emphasis on export-led growth, open markets, and domestic liberalization—while it also provided an opportunity for a response to environmental criticisms of its lending policies. WCED legitimated what former Bank president Lewis Preston called the "win-win" strategy—a phrase that appeared repeatedly in the 1992 report—the Bank adopted. In essence, that strategy meant the "links between efficient income growth and the environment need to be aggressively exploited" (Preston in World Bank 1992:iii).

Other responses, particularly in the North, followed this general interpretation. One important example was a report for the Trilateral Commission, co-authored by the secretary-general of WCED and former OECD environment director Jim MacNeill, which placed the necessity of growth at its core.[6] To make growth sustainable, the authors emphasized the WCED position that environment and economics "must be integrated in all of our major institutions of decision-making—government, industry and the home." Sustainability was defined accordingly, as "the maintenance of a community's or a nation's basic stock of natural capital" (MacNeill, Winsemius, and Yakushiji 1991:20–22). The framing of environmental problems in economic language was typical of post-WCED proposals. According to this view, environmental problems stem from distortions in markets, so solutions require a better application of economic principles:

If nations are to stop depleting their basic stocks of ecological capital, governments will need to reform those public policies that now actively encourage the infamous *des*: *de*forestation, *de*sertification, *de*struction of habitat and species, *de*cline of air and water quality. Virtually all governments today pay lip service to the market, and then they intervene to distort it in ways they find politically convenient. Subsidies, tax abatements, fiscal incentives, price supports, tariffs, and trade quotas of all kinds can distort prices and trading patterns in ways that are economically perverse and encourage unsustainable forms of development. They often rig the market not only against the economy, but also against the environment and, ultimately, against development itself (MacNeill, Winsemius, and Yakushiji 1991:23).

Thus, adapting markets to reflect the cost of natural capital depletion—in other words, getting prices right—should be the basis of development policy to avoid the above distortions.

Domestically, reforms might include an elimination of subsidies in the agricultural, forestry, energy, and transport sectors. These market-distorting measures should be replaced by economic instruments such as environmental taxes to create new market incentives to preserve and enhance natural resources under threat. Other reforms might include an extension of property rights to common resources, that is, to privatize commons such as the atmosphere or oceans. The proposals included tradeable emission permits, water rights, and systems of deposits and refunds on hazardous or recyclable wastes. The Polluter Pays Principle was the guiding norm behind such proposals (MacNeill, Winsemius, and Yakushiji 1991:32–42).

A number of OECD initiatives during this period gave added analytical ammunition and generated political support for the promotion of economic instruments for environmental management. MacNeill's 1984 "Environment and Economics" conference, noted earlier, provided the foundations for later projects on this theme. After 1987, these projects gained greater legitimacy as they were then seen to fit with the thrust of the more widely accepted vision articulated in the Brundtland report. In 1991 the OECD Council endorsed a major project on economic instruments at the behest of its environment committee.[7] The Council proposed, *inter alia*:

- a greater and more consistent use of economic instruments;
- to improve the allocation and efficient use of natural and environmental resources by means of economic instruments to better reflect the social costs of using these resources;
- to seek further agreement at [the] international level on using economic instruments with respect to solving regional or global problems and to ensuring sustainable development.[8]

The OECD report that stemmed from these proposals specifically referred to the Brundtland Commission report as a legitimating source for an interpretation of sustainable development consistent with the recommendations: "The way the notion [sustainable development] was interpreted in the [Brundtland] report implied an enhanced role for environmental economics in actual policy." The OECD report went on to highlight Brundtland's central theme of combining economic and environmental decision-making and the explicit advocacy of economic instruments for sustainable industrial development. In addition, the OECD report noted a number of other conferences and declarations that emphasized the usefulness of economic instruments. They included the Lankawi Declaration on Environment of the Commonwealth Heads of Government (Kuala Lumpur, Oct.

1989), the Bergen Ministerial Declaration on Sustainable Development in the ECE Region (May 1990), the Conference on Environment and Development in Asia and the Pacific (Bangkok, Oct. 1990), the Ministerial Declaration of the Second World Climate Conference (Geneva, Nov. 1990), and the Second World Industry Conference on Environmental Management (Rotterdam, April 1991) (OECD 1994a:13).

As will be shown below, UNCED further cemented this interpretation and the OECD continues to use the Rio Declaration (especially Principle 16) and Agenda 21 chapter 8 as the source of legitimacy for the pursuit of such policies. The OECD report itself went even further than simply endorsing economic instruments, which might include any instrument that affects estimates of the costs and benefits of alternative actions open to economic agents. It favored instruments more consistent with liberal market principles. So, for example, it did not include subsidies in the study since they contravene the Polluter Pays Principle.

Finally, the OECD report argued, dubiously, that the anarchical nature of the international system makes the case for market-based instruments at this level even stronger.[9] In other words, the lack of a world government or strong set of regulatory institutions makes market incentives and instruments more likely to succeed than those that require strict standards and enforcement. Economic instruments may "succeed" not because they necessarily better produce compliance, the report argues, but because they are more likely to even out costs and benefits and provide economic incentives for reluctant parties. Thus the argument for economic instruments was made as much on the basis of efficiency as effectiveness. The report then endorsed international instruments such as emission or energy use charges or taxes, internationally tradeable emission permits, and "joint implementation" programs (OECD 1994a:147–151). The latter refer to a state or company co-financing a project in another state to reduce pollution, then receiving credit for such reductions as part of its own obligations to reduce pollution. As will be shown below, this idea has taken hold in international action on climate change.

The legitimation of these new norms soon became evident in attempts at policy coordination among the Group of Seven (G-7) industrial nations.[10] Although the G-7 at first showed a willingness to accept the Keynesian-style compromise of Brundtland, later it too moved toward a position consistent with MacNeill, Winsemius, and Yakushiji (1991) and the OECD. Interestingly, prior to Brundtland, the G-7 took a position virtually identical to that outlined in the OECD Environment and Economics conference noted earlier. In the Economic Declaration at the 1985 Bonn Summit, G-7 countries agreed that, "We shall harness both the mechanisms of governmental vigilance and the disciplines of the market to solve environmental problems. We shall develop and apply the "polluter pays" principle more widely" ("Bonn Economic Declaration" in Hajnal 1989:293). Later, it went further in support of the more interventionist style of Brundtland, marking at least a formal shift in policy. That shift appeared in 1988 when the G-7 for the first time endorsed the concept of sustainable development at the Toronto summit ("Toronto Economic Summit Declaration: Environment," in Hajnal 1989:372).

It took until 1989 for an interpretation of sustainable development to appear as well as a coordinated response to it. The Paris Economic Declaration appeared to endorse many of Brundtland's proposals and included language that mirrored that in *Our Common Future*. For example, the G-7 declaration used the language of "common goals" to preserve "a healthy and balanced environment in order to meet shared economic and social objectives and to carry out obligations to future generations" ("Paris Economic Declaration," in Hajnal 1989:400). It also voiced a number of the norms mentioned above. For example, it contained statements on the compatibility of economic growth and the environment, the mix of market and regulatory actions, and, significantly for a comparison with later policy, a cautious endorsement of the use of "aid mechanisms and specific transfer of technology" to "help developing countries deal with past damage and to encourage them to take environmentally desirable action" ("Paris Economic Declaration," in Hajnal 1989:401). However, the trend toward integration of environment and economics along liberal market lines was also present. For example, the declaration called for the OECD and United Nations and affiliate organizations to develop techniques to further the use of economic instruments for environmental protection.

By 1990, summit statements had moved toward a stronger support of market mechanisms and away from international aid and domestic regulatory approaches. While environmental problems such as ozone depletion, deforestation, climate change, and marine pollution were recognized, the G-7 argued in Houston that the key to a healthier environment was the recognition that, "strong, growing, market-oriented economies provide the best means for successful environmental protection" ("Houston Economic Declaration," in Hajnal 1991:21). The Houston Declaration did not ignore aid and technology transfer completely, but it gave special emphasis to the OECD's work on environment and economics. It also singled out "market-oriented approaches" as an important area for research on how best to achieve environmental objectives.

Many individual countries have also either sponsored their own programs to formulate international strategies and/or domestic policies that fit with liberal environmentalism or have responded positively to policy proposals along those lines. Two prominent examples are the United States and the United Kingdom. As has been noted, the U.S. administration under Ronald Reagan appeared to see little or no contradiction between environmental protection and the free market. Although actual implementation of such policies was slow (it was not until the Bush administration that serious policy attention turned to market-based incentives for environmental protection), in principle Reagan furthered a trend begun in previous administrations to look for economic incentives and use cost-benefit analysis as guiding principles. The most prominent of such measures came in the Clean Air Act and subsequent refinements in the 1970s, which pioneered the application of air pollution permits to control emissions from U.S. industry. The new Clean Air Act of 1990 expanded this system to include, for example, a permit system for sulfur dioxide emissions that contribute to acid rain.

The impetus for the latter reforms came largely from the "Project 88: Harnessing Market Forces to Protect our Environment" initiative of Senators Timothy Wirth (Democrat) of Colorado and John Heinz (Republican) of Pennsylvania, a project headed by Harvard economist Robert Stavins. The project, which was influenced also by work from the Environmental Defense Fund, argued that market-based incentives provided a cheaper, less intrusive alternative to command-and-control regulation for environmental protection.[11] It was presented at the 1988 Republican Convention in New Orleans, and influenced policy during the Bush presidency.

The people behind Project 88 had strong ties to the Bill Clinton White House. For example, Stavins participated in work for the Progressive Policy Institute (PPI) think tank, a project of the Democratic Leadership Council. Clinton helped to create the Council and headed it from March 1990 to August 1991. The environmental section of PPI's major policy document, *Mandate for Change*, emphasized "free market" ideology, and Stavins detailed there, and elsewhere, specific proposals that all fall under the general rubric of the Polluter Pays Principle (Stavins and Grumbly 1993; Stavins and Whitehead 1992; Hahn and Stavins 1992). Specific proposals from PPI included pollution charges, deposit-refund systems, and tradeable pollution permits. Subsequent reforms within the Environmental Protection Agency, the general policy direction of the Clinton administration, and the public position of Clinton's Vice President Al Gore all suggest these trends became well-entrenched during the Clinton presidency.[12] In addition,

Clinton's appointment in his first term of Wirth to the newly created position of Under Secretary of State for Global Affairs, responsible for international environmental issues, signaled a continuation of strong U.S. support for market-mechanisms in international governance. Wirth, for example, in a second Project 88 report, argued for a tradeable pollution permit system to combat global warming and pushed U.S. policy in this direction (*Project 88—Round II* 1991; author's interview with Wirth).

Similarly, in the United Kingdom, the work of economist David Pearce and his colleagues on market-based mechanisms to promote sustainable development received wide attention (Thomas 1992:73–78). Then Prime Minister Margaret Thatcher's environment minister Chris Patten championed Pearce's ideas and in 1989 Pearce et al. published *Blueprint for a Green Economy*, commissioned by the UK Department of the Environment, which endorsed market-based instruments over traditional standard setting. The arguments used in the study to support his position include the following: such instruments keep down the cost of compliance because the market ensures that those most able to afford to act do so; they act as an irritant to polluters who thus avoid them by creating cleaner technology; and they encourage consumers to choose cleaner products by raising the cost of polluting products.

Pearce remained an influential figure and has published a number of studies that extend his argument from a single economy to the world economy and the developing world in particular. He has argued, for example, that his approach is even more important in the international context because of the potentially huge cost of protecting the global commons (Pearce 1991; Pearce and Warford 1993). Internationally, Pearce and his colleagues' influence can be seen in his reports for institutions such as the World Bank and for the Intergovernmental Panel on Climate Change, the scientific body on which the Framework Convention on Climate Change relies for scientific and policy research (OECD 1994b; IPCC 1995). As one of the lead authors of the report from working group three in the IPCC's 1995 report, on the economic and social implications of climate change, Pearce and his colleagues have had a major influence on the shape of the policy debate on climate change, a development discussed in more detail in later chapters.

A basic assumption that the conditions of property rights are at the root of many environmental problems underlies Pearce's writing. For example, he and his co-authors state in an OECD study on project and policy appraisal that environmental problems commonly arise because the following conditions do not prevail: (a) universality—all resources privately owned and entitlements are completely specified; (b) exclusivity—all benefits and costs of resources accrue to the owner; (c) transferability—owners must be able to transfer property rights to other owners in voluntary exchange; and (d) enforceability—a structure of penalties to keep property from being encroached upon by others.[13] Meeting such conditions, the authors argue, results in win-win solutions to environmental problems. To ensure that capital stocks do not run down (the core of sustainability for Pearce) economic development is an "enabling" condition, thus the core of sustainable development. Creating and enforcing private property rights and using market-based incentives to protect the environment therefore lie at the heart of any strategy for sustainable development for Pearce.

A wide variety of countries initiated, or lent government support to similar programs during the period between Brundtland and Rio, including Australia, Canada, Poland, (then) Czechoslovakia, the former Soviet Union, Belgium, Italy, and a number of other European countries (*Project 88 — Round II* 1991:2–4; Moffatt 1996). These domestic programs combined with the already mentioned EU trend toward liberal market norms give a strong indication of how sustainable development following Brundtland had been interpreted, at least in the North. In the case of the European Community (and then the EU) UNCED reinforced this commitment, as evidenced in the fifth environmental action programme, which places a heavy emphasis on moving from regulatory measures to, "in particular, the greater use of market forces" (Commission of the European Communities 1993:49). It is not surprising, then, that consensus on the direction of international environmental policy was pulled in a similar direction at UNCED. It not only cemented this interpretation of sustainable development, but also gave it international political legitimacy.

UNITED NATIONS CONFERENCE ON ENVIRONMENT AND DEVELOPMENT (UNCED)[14]

The United Nations Conference on Environment and Development arguably evolved as a natural progression from the Brundtland report and domestic and international reforms that followed it. However, the agreements UNCED produced also reflected an inherently political process that Brundtland had more or less avoided. The nature of United Nations multilateral diplomacy ensured that long unresolved tensions left over from Stockholm would resurface at UNCED, which at times appeared to move

UN diplomacy backward to the North-South stalemate of the 1970s. Not surprisingly, the negotiations exhibited some parallels to the Stockholm conference. For example, delegates from North and South disagreed on the degree to which the North, as historically the site of greater pollution, ought to shoulder a greater financial burden for environmental preservation now. They also disagreed on various topics—for example, the relative weight that should be given to issues such as consumption patterns versus population growth, or the need to undertake a broader set of development reforms before the South could be expected to act on global environmental concerns.

Nonetheless, numerous changes since 1972 made a stalemate unlikely. Although tensions remained, delegates overcame many North-South differences to forge a consensus on a relatively well-specified approach to international environmental governance. I have already detailed some of the more important post-Stockholm changes, such as the increased profile of environmental concerns in the United Nations system and other international fora. Changes in the attitudes and understandings among Southern leaders and within Southern coalitions about global environmental problems also made a repeat of Stockholm unlikely. For example, by 1992 most states in the South saw the environment as an important national and international issue: many had national environmental agencies or ministries;[15] saw a link between poverty and environmental degradation; and accepted studies by UNEP, the World Bank and other governmental and nongovernmental organizations (NGOs) that environmental problems such as deforestation, water pollution, or soil erosion harmed the development process (Williams 1993).

These understandings were not confined to elites. By the early 1990s, non-elite opinion in the South also reflected a high level of concern over local, national, and international environmental conditions. Although time series data for the South are not available, a 1992 survey of 24 countries representative of most regions of the world provides a snapshot of attitudes and opinions. Contrary to conventional wisdom at the time, the survey showed little difference between people in wealthy and poorer countries in their concerns over environmental problems.[16] For example, the percentage of respondents who said the environment was a "very serious" problem in their country was 67 percent in Germany, 42 percent in Japan, and 21 percent in Finland (the highest, middle, and lowest ranked countries in the developed world). Respondents in the South gave the same responses in comparable proportions, with 67 percent ranking the environment a "very serious" problem in South Korea, 56 percent in Chile, and 37 percent in the Philippines (the highest, middle, and lowest ranked in the developing world). Answers to a wide variety of other questions indicate similarly comparable patterns in North and South. The responses suggest that environmental issues had penetrated public concern and were considered major issues relative to other core economic and social concerns such as employment or health care. Furthermore, majorities in most developed *and* developing countries were willing to forego some economic growth in order to decrease environmental degradation. On the latter point, pollsters found only small differences between developed and developing countries as a whole. The concern among non-elites in the South was further evidenced by the large-scale participation of Southern NGOs at the Earth Summit and parallel Global Forum.[17]

Although a number of factors likely produced the high levels of public awareness, a series of spectacular international environmental disasters in the 1980s certainly increased public anxiety in the lead-up to UNCED (as they had for Stockholm) and increased pressure for international cooperation and action. The escape of toxic gas at a Union Carbide plant in Bhopal, India (1984), the Chernobyl nuclear accident in the Ukraine (1986), and the Exxon Valdez oil spill (1989) off the coast of Alaska provided vivid examples of how even single environmental disasters could have international repercussions. Big international issues such as ozone depletion had received serious attention by governments, and other global concerns such as tropical deforestation (particularly of rainforests) and biodiversity had also started to gain greater prominence in the public eye. In addition, the hot summer of 1988 in North America galvanized concern over the prospect of climate change and created grassroots momentum that ensured Rio would not just be another UN conference.[18]

As for the other half of the UNCED agenda, 20 years after Stockholm the North did not need convincing that development deserved a prominent place at Rio. Since UNCED originated as a Brundtland Commission proposal, development received equal billing on the agenda from the initial UNGA resolution calling for an international conference on environment *and* development. In addition, the links between environment and development were well established within the United Nations system. Existing institutional arrangements to deal with global environmental problems mostly accepted the linkage and, to varying degrees, had incorporated the linkage into their programs. Although some distrust between North and South remained, the changes since Stockholm meant a Founex would not be required: unlike in 1972, the conference secretariat did not need to devote time and energy to convince developing countries to participate or that the global environment was an issue worthy of an international response. Rather, substantive negotiations focused on the division of responsibilities, rights, and obligations in regard to global environmental action, the means of taking action and type of action required, and the source of financial and technical resources to make action possible.

The larger political context of the Cold War's end also created an opportunity for environment and development issues to get a serious hearing. A new optimism prevailed around the ability of states to cooperate to solve global problems previously unable to compete for attention on the international diplomatic agenda. The combination of environment and development (and perhaps democracy and human rights which Rio addressed only tangentially) epitomized the alternative international agenda so long buried under the preoccupation with superpower conflict. Rio represented not only an airing of those concerns, but a chance to show the new face of multilateral diplomacy and global cooperation. An open, market-friendly international economic system and a peaceful, multilateral political system were to be the cornerstones of the post–Cold War international order.

This context meant the organizers of UNCED saw in it an opportunity to make a fundamental statement on global governance, not just concerning the environment, but on how planetary affairs ought to be managed. Whereas social welfare and human rights summits of the 1990s, important as they were, seemed aimed at promoting fairer governance within states and setting universal standards and programs, the linkage of environment and development—and the concept of sustainable development that promoted that linkage—seemed the most direct challenge to human activity on a global scale and a fine focus for the new global order. It addressed the core challenge to the international political economy as rich or poor, North or South, strong or weak would have to face the same repercussions.

The Earth Summit even put forward a new notion of planetary security. From his opening speech at the first PrepCom, Maurice Strong stated the linkage clearly:

> People and nations have always been willing to accord highest priority to meeting threats to their security. In this case the security of our planet and our species is at risk. Surely this must be seen as the ultimate security risk which calls for the ultimate security alliance (quoted in Speth 1990:41).

World leaders, policymakers and academics—including former Soviet leader Mikhail Gorbachev and former U.S. Vice President Al Gore—have since promoted this broadened notion of security that became popular immediately following the end of the Cold War.[19] These factors combined to elevate UNCED's importance in a way that may seem naively optimistic today, but ensured from the start that the Earth Summit would be much more than an environmental conference like Stockholm.

To say, then, that UNCED resulted merely from a North-South compromise misses this political and economic context. It also misses twenty years of "learning" within international institutions, governments, and societal groups. From these changes in practices and discourses around the environment and development, the final compromises drew their substance, legitimacy, and support. Thus the results of the Earth Summit were both evolutionary and revolutionary: they evolved from ideas most clearly voiced in the Brundtland Commission and were forged by a political process that reproduced that learning process at the level of governmental negotiations. The end point reached, however, appeared revolutionary to the degree that it finally entrenched the shift in norms from a juxtaposition of environmental protection and development to the compromise of liberal environmentalism.

Much has already been written about the Earth Summit from a variety of perspectives.[20] Hence I will not try to summarize the proceedings, which involved thousands of official delegates from governments and NGOs, thousands of additional NGOs from a variety of backgrounds at the parallel Global Forum, and a huge and wide-ranging agenda that took shape over two and half years and dozens if not hundreds of official and unofficial gatherings from the time the United Nations called for a conference in 1989. As in the earlier discussion of Stockholm in the previous chapter, below I concentrate closely on the official preparations and negotiations and look mainly at how ideas eventually meshed into the normative framework—the norm-complex—agreed to at Rio. The most attention will be paid to negotiations over the Rio Declaration and Agenda 21, although some reference will be made to other treaty negotiations where delegates hashed out some core issues. Subsequent chapters will examine in more detail the source of ideas that dominated UNCED and why those ideas became institutionalized as norms.

The Conference and Normative Context

The Earth Summit, held June 3–14, 1992, brought together 178 states (more than 100 of those represented by heads of state or government), 1,420 accredited NGOs[21] at the conference, and another 8,000 NGOs at the Global Forum, held nearby to coincide with the official conference.[22] Major conference outcomes included the Rio Declaration on Environment and Development, the detailed 40-chapter action plan of Agenda 21, and the nonbinding statement of Forest Principles.[23]

Two major environmental treaties were also opened for signature at Rio, but negotiated in separate processes. The Framework Convention on Climate Change (FCCC) was negotiated by an Intergovernmental Negotiating Committee established by a resolution of the UNGA beginning in 1990. The Convention on Biological Diversity was negotiated starting in 1989 by an *ad hoc* working group of experts mandated by UNEP's governing council, although negotiations were open to states not on the governing council. In 1991 the negotiating group was renamed the Intergovernmental Negotiating Committee. UNCED also established a new institution, the UN Commission on Sustainable Development, to oversee the implementation of Agenda 21.

The proposal for a global conference on environment and development came directly from a recommendation by the Brundtland Commission. Thus, the December 22, 1989 General Assembly resolution 44/228 calling for a global conference explicitly linked environment and development under the concept of sustainable development. Not surprisingly, the resolution itself contained some vague wording that stemmed from uneasy compromises between North and South and those conflicts pervaded much of the conference process. For example, countries from the North primarily pushed for a global conference on the environment to coincide with the 20th anniversary of Stockholm, while many countries from the South feared that such a conference would have a strong environmental (Northern) bias and not focus enough on development concerns (Chasek 1994b:46). Nonetheless, a year after the General Assembly first considered the idea, states agreed on resolution 44/228, in effect recognizing that environment and development had become inexorably linked when it came to addressing environmental problems on a global scale. The final wording thus called for a global conference that "should elaborate strategies and measures to halt and reverse the effects of environmental degradation in the context of increased national and international efforts to promote sustainable and environmentally sound development in all countries" (United Nations 1989). In terms of the evolution of environmental governance, the question the conference would answer was what formulation of sustainable development would prevail.

One sign of that direction was the absence in conference outcomes of the qualifier "environmentally sound" that appeared in resolution 44/228.

According to Pallemaerts (1994:15) the modifier was added in the first place because the remainder of the resolution largely supported the status quo of the international economic system, thus supported economic growth as the major concern. He argues that the drafters of the resolution were not convinced that ecological concerns would automatically be incorporated by the concept of sustainable development unless texts explicitly recognized their importance. Those concerns proved prescient, as UNCED outcomes were more definite on the promotion of a liberal and growth-oriented economic order and less so on ensuring ecological viability. The form of governance that emerged from UNCED emphasized one particular pathway from the concept of "sustainable development" to produce a set of norms that legitimated the compatibility of liberal economics and environmental protection. The formulation in the Brundtland Commission report did not determine this path of governance outright, but its emphasis on growth legitimated the linkage of environmental concern to liberal economics and helped de-legitimate forms of governance that might be seen in opposition to leading economic principles that did encourage growth.

Whereas the post–Cold War political context probably facilitated cooperation generally, the shift in international economic governance toward the liberal orthodoxy of the "Washington Consensus," and its widespread support, influenced the direction that cooperation was likely to take. In contrast to analysts who contend that the market, reinforced by this post–Cold War triumph of liberal market based economics, marks a challenge to environmental governance, I argue that UNCED embraced and even anticipated the new orthodoxy in its formulation of norms of international environmental governance (Haas 1996:43–44).

To take one important example of the normative shift, the decline in legitimacy of the "Common Heritage of Mankind" principle (CHP) can be contrasted to the successful entrenchment of the Polluter Pays Principle (PPP) by UNCED.[24] The former proposed that areas not under any state's jurisdiction be subject to common property ownership and shared economic use.[25] It originally gained prominence in the negotiations for the 1982 UN Convention on the Law of the Sea (UNCLOS III), but also appeared in slightly altered form in the Outer Space Treaty of 1967 and Moon Treaty of 1979.[26] However, by 1992 it had fallen out of favor in international fora that addressed problems of regulating the global commons and environmental issues in general.

As an illustration, a complete search of UNCED documents reveals that CHP did not appear in any of the agreements reached, not even in Agenda

21. The CHP was mentioned briefly in opening or closing statements of only 10 states (of 178 that attended) or international organizations and a handful of regional reports. Of those, only three states (Portugal, Kenya, and Jamaica) mentioned its specific application, referring to the Law of the Sea and Outer Space treaties, while other specific references to it were by developing states who *did not* want it applied to biodiversity.[27] In particular, the CHP met a hostile reception by developing countries in negotiations on forestry and biodiversity, especially because they argued it infringed sovereignty (Imber 1994:57–63). This marks a departure from consensus on CHP in the World Conservation Strategy, which states that gene pools "are the common heritage of mankind" (IUCN 1980: section 12.1), and in a major Food and Agricultural Organization (1983) statement on genetic resources (see also Mensah 1994:47). Northern countries distanced themselves from the concept because they associated it with a general program of global economic management and redistribution and in opposition to market-based principles. Significantly a new implementation agreement for UNCLOS, adopted by the UNGA in July 1994 and signed by formerly recalcitrant states including the United States, effectively altered the meaning of CHP so relevant portions of UNCLOS (that is, Part XI on deep sea-bed mining) would conform with market-based principles.[28]

As a result of this shift, many states at UNCED used the language of areas or issues of "common concern" but refused to invoke CHP. This new language took over from the CHP in major agreements as well. For example, the Convention on Biological Diversity "*affirms*" in the preamble that, "the conservation of biological diversity is a common concern of humankind" while "*reaffirming* that states have sovereign rights over their own biological resources." Operationally, access to genetic resources under the convention (Article 15) moves away from the common heritage norm found in the earlier FAO (1983) Undertaking on Plant Genetic Resources and entrenches the "sovereign rights of States over their natural resources" and national governments' legislative "authority to determine access to genetic resources."

The debate over climate change showed a similar pattern. When governments first raised the issue in the General Assembly in 1988, Malta, which originally proposed CHP in UNCLOS III negotiations more than twenty years earlier, requested the inclusion of an agenda item entitled "Declaration proclaiming climate as part of the common heritage of mankind." However, support for the concept quickly eroded as it became clear that climate change might actually receive serious international attention. When the General Assembly endorsed the creation of the Intergovernmental Panel on Climate Change (IPCC) later that year, CHP was out. Instead, the UNGA resolution was amended to refer to climate as the "common concern of mankind," and CHP never again received serious consideration in relation to climate change (Bodansky 1994:52).

In contrast, PPP, introduced into international discussions at about the same time as CHP, started to gain prominence in the late 1980s after its support in the Brundtland Commission. It can now be found in a wide range of international agreements and programs including Principle 16 of the Rio Declaration, article 130R of the Single European Act, EC/EU programs and legal instruments, and OECD Council Recommendations.[29] At least one scholar argues it has the status of a general principle of international law, and most acknowledge that at least among OECD countries and within the EU it is recognized as a customary rule of international law.[30] The vast majority of states at UNCED also endorsed PPP both nationally and internationally in their statements and reports, and most of those claimed to have implemented it at the national level to varying degrees (IDRC 1993).

As explained in the previous chapter, the OECD intended PPP not as a rule of liability, but as a means to avoid environmental regulations that might alter the operation of the market and particularly of free trade. It aims to internalize environmental costs to ensure continued economic growth by minimizing trade-offs between economic efficiency and environmental protection. Implementation of PPP demonstrates the trend in international environmental institutions to move toward market-based solutions to environmental problems consistent with the principle. Moreover, the growth-oriented ideology behind the PPP has clearly found its way into a wide range of international statements and agreements and constitutes a dominant meaning of sustainable development.

The Common Heritage principle may not be completely dead,[31] nor is PPP universally implemented.[32] The argument rather highlights CHP's low level of institutionalization consistent with its original meaning, and the poor prospects for common ownership schemes to form the basis of attempts to manage global environmental problems, in contrast to the much greater legitimacy enjoyed by PPP.

The Negotiations

The negotiating process for Rio had a number of similarities to Stockholm. Most obviously, Maurice Strong was picked again as secretary-general (he had also been a member of the Brundtland Commission). The literature on UNCED also singles out Tommy T.B. Koh of Singapore, elected chairman of the Preparatory Committee (PrepCom), as a key leader who played an equally important role in moving delegates forward on divisive issues (e.g., Spector et al. 1994). Koh had also worked with Strong in the preparations for Stockholm and had served as president of the UN Conference on the Law of the Sea in 1981 and 1982. The UNCED secretariat and bureau also generally played important leadership roles in the preparation process, as great power leadership (especially from the United States) was lacking, especially in the early going.[33]

Also like Stockholm, four PrepComs preceded the conference—one in Kenya (March 1990), two in Geneva (March/April and August/September 1991), and one in New York (March/April 1992). Procedural and organizational wrangling pushed most of the substantive issues to the final PrepCom in New York.[34] The pattern of slow progress changed as the conference date approached and the prospect of failure grew. Changes in the selection of delegates reflected the increased political stakes, as technical experts that had dominated earlier meetings were supplemented or replaced by political strategists with experience in multilateral diplomacy.

The character of the New York session differed in process as well. The pace of negotiations picked up with more late-night (and all-night) meetings, closed-door gatherings of small informal contact groups of states, and less formal meetings of working groups and plenaries (which meant far less NGO access). Not all issues could be resolved in the short time period that remained before the conference and a number of the most acrimonious points were left bracketed (that is, with disagreements left in the text that required further negotiation) in the texts sent on to Rio. In addition, a number of specific proposals, such as those related to atmospheric issues and biodiversity, were discussed late or not at all since relevant issues remained unresolved in the parallel negotiations on climate change and biodiversity. Negotiations on financial resources also broke down on the last day of PrepCom IV despite being given the highest priority. Nonetheless, delegates reached agreement on 85 percent of Agenda 21, although the remaining 15 percent contained many of the toughest issues and had to be negotiated during the conference itself.

In negotiations, developing countries initially tried to forge a unified position and negotiate as the traditional G-77 plus China bloc. Some in this group hoped that the environment could be a new bargaining chip to reassert a Third World coalition weakened by the debt-ridden 1980s and the failure of the NIEO (Williams 1993). Apart from seeking specific interests in texts on sectoral issues (for example, forests, energy, and hazardous waste), the G-77 focused on four main principles:[35]

1. New and additional development assistance and equal say for developing countries in decision making.
2. Reduction in consumption of natural resources and environmental services in the North to give the South "environmental space" for its development.
3. No restrictions on imports to industrialized countries on environmental grounds.
4. Technology transfer on preferential and concessional terms.

The one big success of this strategy was to entrench the idea of "common but differentiated responsibility" of states to protect the global environment. This principle can be found in the FCCC and the Rio Declaration and its acceptance ensured that some equity considerations would guide international policy. However, the larger hopes of developing countries to secure substantial new financing, or use the environment/development nexus to change international economic norms, were never realized, nor does it appear that many Southern states fought hard to fundamentally change economic norms, as they had in previous global negotiations (Porter and Brown 1991:117). Rather, the general thrust to support a right of development (Principle 3 of the Rio Declaration) and related development norms were generally phrased in such a way as to be compatible with current liberal economic norms, while states agreed on basic environmental concerns embodied in such new norms as the Precautionary Principle (Principle 15 of the Rio Declaration) with relatively little difficulty.[36]

Furthermore, even the latter norm could be interpreted as fully compatible with liberal environmentalism. The principle essentially argues that in the face of uncertainty, action is still warranted under conditions of high risk of potentially severe environmental damage. It fits the use of market instruments that aim to prevent waste generation at the source by incorporating costs up front rather than by means of end-of-pipe regulation.[37]

These latter outcomes did not necessarily go against the South's interests, but reflected a slightly different reality of North-South relations than implied by the apparently unified position found in documents such as the South Centre's (1991) report on environment and development. For example, the controversy over "additionality" reveals how traditional G-77 goals became conflated with specific objectives in the UNCED negotiations (Jordan 1994a). Developing countries argued that the North, as the historical site of the majority of global pollution and the source of environmental damage, ought to help pay for the costs of environmental measures taken in developing countries (that the North desired). According to the principle of "additionality," any such money ought to be new and in addition to monies already committed for North-South aid.

Ozone negotiations set some precedent for the norm since developing countries received a commitment for new and additional monies as part of the 1990 (London) amendments to the Montreal Protocol on Substances that Deplete the Ozone Layer. The parties set up a Multilateral Ozone Fund to assist developing countries, especially India and China, which were holdouts to the initial agreements (Parson and Greene 1995:20). The G-77 wanted similar mechanisms in other major treaties, such as climate change and biodiversity, but achieved only limited success. The language of those agreements (and even the ozone agreement) carefully avoided a commitment to the norm of additionality or the suggestion that additional funds for a particular environmental problem ought to set a precedent for responses to other problems. For example, developed countries did not quantify their commitment to provide additional resources at concessional levels (or grants) to meet the "incremental costs" of developing countries to enable them to comply with the treaties (Jordan 1994a:28). Furthermore, in each case, the GEF now manages the funds, which suggests the underlying conditions for the arrangements are unlikely to stray far from the liberal economic norms supported by the World Bank.

"Additionality" became so controversial not simply because of dwindling aid budgets, but also because many developing countries saw the debate as a way to revive the more radical goals of institutional or economic restructuring reminiscent of the NIEO. Other developing states sought special consideration but did not oppose the normative thrust of Northern proposals. The least developed and/or most debt-ridden countries also showed reluctance to advance a broader normative agenda since they felt more vulnerable than in the 1970s and were weary of antagonizing industrialized countries (Porter and Brown 1996:117). Despite the strongly stated position for additionality by developing countries in negotiations, unremitting opposition voiced most forcefully by the United States prevented the discussion progressing to consider specific commitments or discuss in detail how to gain additional financial resources. It also appeared that some developing countries that saw the larger potential benefits of the conference were not willing to allow negotiations as a whole to breakdown on this issue. Meanwhile, while some developed countries voiced some support for the idea of additionality in principle, others such as Canada and the United Kingdom, already facing dwindling budgetary resources for ODA, quietly allowed the United States to take a hard-line position with their tacit approval, while staying in the background on this issue or putting a more positive spin on what came out essentially to the same position (Strong 2000:208; Ricupero 1993).

The ambiguity in the nature of the goal of additionality made it more likely that countries such as the United States, which had opposed the notion since Stockholm, would regard proposals as part of a broader agenda for institutional reform. Instead, the United States argued at UNCED's third preparatory meeting that sustainable development could adequately be paid for by utilizing existing resources more efficiently and by drawing on the private sector (Jordan 1994a:19). The efficiency that would be gained by reliance on the private sector, for example, or the Polluter Pays Principle would produce the needed additional resources. In the end, a G-77 proposal at the final PrepCom, put forward by Jamsheed Marker of Pakistan, finessed the issue by accepting an acknowledgement in principle that new and additional money would be forthcoming without insisting on specific commitments or mechanisms (Strong 2000:213). In this watered down form, Brazil's Rubens Ricupero,[38] the coordinator of UNCED's contact group on finances, was able, during conference negotiations themselves, to engineer a compromise for the financial chapter (chapter 33) of Agenda 21.[39] It acknowledged that Agenda 21 required "new and additional" financing for developing countries to be implemented, but contained no specific commitments to provide it. The norm did not appear at all in the Rio Declaration.

The norm of common but differentiated responsibility avoided the ambiguity or divisions, even if minor, among developing countries that made it easier for resistant developed countries to block consensus on additionality. Whereas the norm of common but differentiated responsibility supported the idea that different levels of environmental protection might be expected of rich and poor, or grace periods might be allowed for costly domestic reforms,[40] it implied less about changes to governing international institutions or the need to reshape the international political or economic order.

The outcomes of UNCED reflected this more modest goal. For example, money pledged at UNCED would be financed primarily through the GEF, which reflected World Bank policy and norms—although developing countries later gained more say in GEF governance; developing countries achieved no real concessions on technology transfer which remained mainly through commercial means; and OECD countries in their statements and actions often predicated concessional financing (a primary condition for additionality sought by G-77) on market and policy reform (Haas, Levy and Parson 1992; Jordan 1994a:19–20). Even the GEF, which appears now to be a permanent institution, does not solely represent "additional" funds, but often money diverted from other development assistance programs at the discretion of donor countries. Whenever additional finances are mentioned in UNCED documents, such as Agenda 21, the language is vague, avoids specific monetary goals or mechanisms, and does not generally differentiate between resources to be committed for environmental or more traditional development purposes. The compromise wording on development aid states that countries would "reaffirm" their commitment to reach the UN target of 0.7 percent of GDP for official development assistance and augment aid programs to reach that target "as soon as possible." In general, the downward trend in development financing from North to South that had already begun by 1992 continued rather than being altered by UNCED, with aid levels in the 1990s averaging just under half the 0.7 percent GDP target sought by developing countries (Jordan 1994a:26–27).

So whereas some authors suggest the environment, and UNCED specifically, provided a renewed opportunity for a Third World coalition, the different objectives and concerns of many developing countries, not least of which being the economic and ideological differences between them, prevented any kind of push for a radical normative agenda like the NIEO. Negotiations on a number of specific issues also did not break down along North-South lines. The G-77 provided draft texts for all the UNCED negotiations, but states within the coalition often divided into smaller coalitions on issues of direct interest to them. For example, coalitions formed around states with highly fragile mountain ecosystems and among a group of small low-lying island states likely to be most affected by rising sea levels caused by global warming. The negotiations over climate change in particular caused rifts in the G-77—with small island states and oil producing states taking opposite positions—that continued to grow after UNCED.[41] These specific splits did not generally affect negotiations on basic norms, however (Sjöstedt et al. 1994:17; Williams 1993).

Industrialized countries also split on a number of issues. The United States was the least sympathetic to developing country concerns, particularly if they appeared to threaten U.S. freedom of economic action. As a result, the United States did not play a leadership role, and the Bush administration appeared disengaged until very late in the negotiation. At that point, it used its clout more to block initiatives it disagreed with (such as targets and timetables on limiting Greenhouse Gas emissions), rather than to propose compromises or push for more far-reaching agreements (Porter and Brown 1996:118; Hajost 1994). The Nordic countries showed greater sympathy to developing country demands and the EU as a whole fell somewhere in between, as did Japan (Porter and Brown 1996:118). Different countries took the lead (or acted as spoilers) on particular policy initiatives, but a specific negotiating strategy in the North was not apparent, perhaps owing to the lack of U.S. leadership and splits within the EU.

In terms of the normative development, the underlying emphasis on market norms, even when combined with developing country demands on issues such as a "right to development," meant North and South were really not as far apart on core issues as some accounts have argued. For example, Porter and Brown (1996:120) point out that despite some reluctance from both developing and some developed countries, states reached agreement in the negotiations on Agenda 21 to remove or reduce subsidies inconsistent with sustainable development (such as sales of timber from public lands at below costs of production) and to improve price signals through environmental charges or taxes. Similarly, Malaysia, one of the developing countries most opposed to the "eco-imperialism" of the North in forestry negotiations, used among the strongest market-led, right to growth rhetoric (Imber 1994:98). Market-friendly measures were supported in the Rio Declaration and a number of specific proposals in chapter 8 of Agenda 21, on Integrating Environment and Development in Decision-Making. Meanwhile, any move to alter the international liberal economic order (such as support for commodity price agreements) was opposed by the United States and other OECD countries.

The Rio Declaration and Norm-Complex

The Rio Declaration on Environment and Development articulated and legitimated the trend in environmental governance toward liberal environmentalism, and best reflects currently prevailing norms defined as collectively held views of appropriate behavior. Although it only constitutes "soft law" like its predecessor at Stockholm, the Rio Declaration "is the one 'product' of UNCED designed precisely to embody rules and principles of a general and universal nature to govern the future conduct and cooperation of States" (Pallemaerts 1994:1). It reflected "to the extent any international instrument can do so—the current consensus of values and priorities in environment and development" (Porras 1994:20).[42] The other UNCED outcomes, especially Agenda 21, but also the biodiversity and climate change treaties, reflected the norms in the Declaration.

The Rio Declaration is arguably a more ambiguous document than its Stockholm predecessor and contains obvious political compromises and some vague language. Nonetheless, its preamble and 27 principles demonstrate a much greater synthesis of the environment-development nexus than did the Stockholm Declaration. Those who see it as a step backward from Stockholm most often point to its more anthropocentric focus, its further entrenchment of state sovereignty, and less attention to concrete environmental or conservational concerns. That is all true. But the Declaration is not a failure from the perspective of the synthesis that the organizers of Rio sought or of how environmental governance had in fact evolved. To the contrary, the Declaration provides an accurate "snapshot of history" of what I argue was the emerging normative consensus of liberal environmentalism (K. Thompson 1993:85). As Marc Pallemaerts put it, within the Rio Declaration, "[T]he liberal economic order . . . acquires for the first time a normative character in an international instrument relating to the environment, as States commit themselves to 'promote' this system in order to 'better address the problems of environmental degradation'"(1996:633–634).[43]

Negotiations for the Rio Declaration got off to a slow start owing to organizational and procedural wrangling during the first two PrepComs in 1990 and early 1991. The working group that would negotiate the Declaration (Working Group III on legal and institutional issues) was not established until PrepCom II when delegates agreed it would prepare what Maurice Strong initially hoped would be an "Earth Charter." Reminiscent of his early goals for Stockholm, Strong envisaged an inspirational statement of care for how nations and people ought to treat the Earth and one another. In PrepCom III, however, developing countries made it clear they would not accept a document that seemed destined to focus too heavily on environmental concerns (Grubb et al. 1993:85; Chasek 1994b; Chatterjee and Finger 1994:49). Neither could much support be found for a legal document like the Universal Declaration of Human Rights, the original vision of a statement of principles proposed in the Brundtland report. A group of legal experts commissioned by Brundtland hoped such a declaration might later evolve into protocols with specific rights and obligations (WCED 1987:332). However, the project of codifying international environmental law, started by this renowned group of experts from North and South, never got a serious hearing.[44] Not surprisingly, then, an early Canadian draft proposal of a legal statement of rights and obligations (reminiscent of a similar Canadian proposal at Stockholm) garnered little support (Palle-maerts 1996:627–629; Shibata 1994:33–34). As a result of these conflicts, the discussions in Working Group III revolved around what the statement of rights and principles should be called, with G-77 countries insisting the title better reflect development concerns. Eventually, a Malaysian proposal to call the document the Rio Declaration on Environment and Development won out over Strong's Earth Charter.

Negotiations over the substance of the statement of principles (as with nearly all substantive issues) took place almost entirely in PrepCom IV. The working group assigned to the task started the five-week session with a compilation text of more than 136 paragraphs. That got whittled down and massaged into the 27 principles of the Rio Declaration, which emerged from the session as the only unbracketed document sent on to Rio. The debate itself was based largely on a G-77 draft text. PrepCom chair Tommy Koh had to step in a number of times throughout the session to work out compromises between G-77 and other parties, and eventually set up a new drafting committee of eight G-77 and eight OECD delegates two days before the end of the session. Although a number of delegations were unhappy with Koh's methods and various aspects of the final text, he successfully used his negotiating skill and personal and political capital to forge an acceptable normative consensus that synthesized proposals from a variety of draft texts (Chasek 1994b:56).

The stalemate in negotiation came not over how to conceive of the environment/development nexus, but on how rights and obligations ought to be divided between North and South. The South wanted emphasis on state sovereignty and an increased obligation for environmental protection to fall on the North, while the North wanted a more equal burden-sharing closer to the common responsibility approach of the Brundtland Commission. The draft proposal submitted by G-77 articulated their general goals listed above. The most important of the specific norms proposed included state sovereignty, common but differentiated responsibility, a right to development, no use of environmental considerations to justify trade restrictions, and a right to adequate "environmental space" for developing countries to allow as much room to develop as the North had required (South Centre 1991; Porter and Brown 1994:126; Mensah 1994). Put bluntly, environmental space meant space to pollute in order to develop. This norm would have fit with the norm of common but differentiated responsibility in that the North would have been obligated to reduce emissions, change patterns of consumption or production, develop new technologies, and so on, first and to a greater degree than the South. That way, the South would have an equitable opportunity to pollute as compared to the opportunity the North had historically enjoyed. However, the South eventually dropped this norm as it was the most unacceptable to OECD countries. The other norms listed above did appear in the final draft in one form or another.

The strong position of the South put some countries in the North, particularly the United States, which came into the negotiations with a resistance to any new commitments, on the defensive. The United States, for example, tried to block any wording that implied specific responsibilities. Northern countries also deleted a principle proposed in Koh's compromise draft that identified industrialized countries' consumption patterns as the "main cause" of environmental degradation, and another one that would have entrenched additionality and technology transfer on preferential and concessional terms (Porter and Brown 1994:127). These conflicts related much more to state responsibility than to either development norms themselves or the basic compromise of liberal environmentalism, which placed environmental protection as firmly fitted within a liberal economic system. Thus the contributions from the North that promoted liberal environmentalism easily found acceptance. For example, the parts of the U.S. draft that promoted open and free markets (but that markets should also reflect full economic accounting of environmental costs and benefits) and the Polluter Pays Principle made it into Koh's final synthesis and appeared to cause little disagreement (Grubb et al. 1993:86).

It is arguably significant that the Rio Declaration emerged from the PrepCom in its final form while other documents did not. It demonstrated that a normative consensus was largely present going into the Rio process, although a number of specific formulations had yet to be resolved. Negotiations did not require the same kind of trade-offs among various interested parties that characterized negotiations on a number of specific environment and development problems addressed in Agenda 21, for example. In this sense, UNCED was indeed successful in institutionalizing a legitimate norm-complex—or as others have called it, a regime of sustainable development or new international law of sustainable development (Spector et al. 1994; Sands 1993; Pallemaerts 1996)—even if some environmentalists were unhappy with the result (see Chatterjee and Finger 1994; Sachs 1993; Pallemaerts 1994, 1996). While UNCED might be criticized for not producing enough concrete action on particular issues, it did achieve the institutionalization of a particular vision or understanding of how the international community ought to manage or approach global environmental problems and the norms that would guide future action.

The principles themselves are not easily grouped as many combine elements of environment and development. Below I will highlight key principles as they demonstrate changes in norms or entrenchment of norms already present since Stockholm or the Brundtland Commission report. In general, the norm-complex of liberal environmentalism articulated in the Rio Declaration supports sustained economic growth, free trade, privatization of the commons, and the use of market-based instruments as the preferred means of environmental protection.

The Declaration starts with a human-centered vision of the environment, stating in Principle 1 that, "Human beings are at the center of concerns for sustainable development." Human beings should live "in harmony" with nature, but the anthropocentric focus is striking in comparison to earlier global declarations. Although the Stockholm Declaration and 1982 World Charter for Nature viewed human beings as primary, they clearly recognized ecological limits and the inherent value of the natural environment and other species. The Rio Declaration does not delineate the various aspects of the environment that require protection or management as did the first seven principles of the Stockholm Declaration. Rather, the core norms around sovereignty and the importance of human-centered development are dealt with immediately.

Principle 2 on sovereignty reproduces almost verbatim Principle 21 of the Stockholm Declaration, but adds that states have the sovereign right to exploit their own resources "pursuant to their own environmental *and developmental* policies" (emphasis added). Reading the Declaration as a whole makes clear that developmental policies mean liberal economic and growth-oriented policies, and that environmental concerns ought not to limit a state's ability to pursue such policies by, for example, imposing trade restrictions based on environmental concerns. The point is made explicit in Principle 12:

> States should cooperate to promote a supportive and open international economic system that would lead to economic growth and sustainable development in all countries, to better address the problems of environmental degradation. Trade policy measures for environmental purposes should not constitute a means of arbitrary or unjustifiable discrimination or a disguised restriction on international trade. . . .

Clearly free trade and environmental protection are seen as compatible under this formulation. Arguably, free trade and liberal economic policy more generally are viewed as *necessary* for successful environmental protection. Indeed, Principle 12 reproduces, almost verbatim, sections of GATT article XX (on public health and safety exceptions to general obligations not to raise trade restrictions) that have been used in practice to limit environmental restrictions on trade.

The sovereignty provision mostly articulates what states already recognized as the basis of international environmental law since Stockholm. But Rio helped to further entrench state sovereignty by incorporating the norm in Principle 2 into the other Rio agreements. For example, the debate on forests during the PrepComs became polarized between states, such as the United States and Canada, that argued for a "global responsibility" approach, and Malaysia and India, who argued for "sovereign discretion." Malaysia, India, and other developing countries feared the former approach would lead to forests being viewed as part of the "Common Heritage" norm, which, as I pointed out earlier, they strongly opposed on the grounds that it would potentially allow Northern states to unduly influence decisions on forests within the jurisdiction of Southern states (Porter and Brown 1996:126). Similarly the climate change treaty incorporates the newer Rio interpretation of state sovereignty, which by emphasizing development as much as environment, further reinforces that development policies ought not to be interfered with on environmental grounds. Thus the preamble to the FCCC recalls the "pertinent provisions" of the Stockholm Declaration rather than identifying Principle 21 directly, and then reproduces verbatim the language of the Rio Declaration's Principle 2. Biodiversity actually uses the sovereignty language of Stockholm Principle 21, but its substantive provisions reinforce the newer interpretation by not imposing limits on environment or development policies that may affect the environment beyond the limits of national jurisdiction (Pallemaerts 1994:7).

The other side of sovereignty—state responsibility for activities that cause environmental damage to other states or common areas—is also present. However, only minor progress had occurred prior to Rio on liability for environmental damage in international law, and Rio did little to advance this area of law. Principle 13 merely exhorts states to develop "national" law regarding liability, and limits the development of international law to liability and compensation for "adverse effects of environmental damage caused by activities within their jurisdiction or control," a narrower formulation than Stockholm's Principle 22 (Pallemaerts 1996:639–640; Kiss 1994:60). Similarly, Principle 14 calls on states to "cooperate" to prevent the relocation of "activities or substances that cause severe environmental degradation or are found to be harmful to human health." It also does not set up any liability; rather it is a minor acknowledgement of developing country concerns that they not be the recipients of unwanted hazardous waste exports.[45]

Only recently have states shown some willingness to develop specific liability rules in international environmental agreements. Most notably, parties to the 1989 Basel Convention on the Transboundary Movement of Hazardous Waste adopted a Protocol on Liability and Compensation on December 10, 1999, making it the first major multilateral environmental treaty to include a liability and compensation regime. The Protocol refers specifically to Principle 13 in preamble as its normative basis.[46] Interestingly, even this agreement, which seems to be the exception to the general thrust of liberal environmentalism, was adopted together with a Ministerial Declaration on Environmentally Sound Management that attempted to steer the management of hazardous wastes in ways more consistent with the norm-complex institutionalized at Rio. For example, the Ministerial Declaration reiterates a commitment to the implementation of the Rio Declaration and Agenda 21, and recognizes the need to focus efforts at prevention of waste at source rather than rely on compensation and liability. The emphasis on prevention is more consistent with PPP and cost internalization. Indeed, the declaration explicitly recognizes "the need to develop strategies that will harness market forces to promote waste minimization and environmentally sound management." In order to achieve sound management, the parties decided to support a program that promoted "financial and other economic instruments or concepts, with a view to identifying sustainable and self-sufficient solutions for the minimization and environmentally sound and efficient management of hazardous and other wastes subject to the Basel Convention." Management norms consistent with liberal environmentalism have thus informed the evolution of the Convention since Rio.[47]

One advance on state responsibility from Stockholm was that Principle 18 obligates states to notify others of natural disasters or other emergencies "likely to produce sudden harmful effects on the environment of those states." Likewise, Principle 19 obligates states to give advance notification about activities that are likely to cause environmental damage in other states, and to consult with those states. These principles have been entrenched in a number of other treaties and declarations since 1972 when states could not agree on the norm (Kiss 1994:59–60).

A number of the principles articulate norms that I have grouped under the heading of the political economy of environment and development. Principle 3 proposes the controversial "right of development," which had been strongly opposed by the United States during negotiations.[48] Like the term sustainable development itself, the Declaration never defines development. Yet, significantly, the "right to development" appears before any mention of "sustainable development." In other words, traditional development goals should not be inhibited by "sustainable development," if defined any differently than development in its classical sense (Pallemaerts 1996:632). In line with the general thrust of the Rio agreements, the WCED, and other development norms, development in this context appears to mean mainly the promotion of economic growth.

Principle 4 articulates the most general statement of how environment and development are to be linked. It states: "In order to achieve sustainable development, environmental protection shall constitute an integral part of the development process and cannot be considered in isolation from it." The two-way relationship implies both the need for policies such as environmental assessments (Principle 17)[49] and that environmental concerns ought to fit into overall strategies for development. The precise way in which environmental policies ought to be integrated into economic policy must be interpreted through other parts of the Declaration and other legal instruments and policies. The EC/EU was and remains the most advanced jurisdiction in integrating environment and economics, thus practice there is one indication of the implications of this norm. As argued above, EU policy has generally moved in the direction of supporting norms consistent with fitting environmental protection into a liberal economic system that promotes growth (Sands 1994:xlv–xlvi).

The promotion of an open (liberal) international trading system in Principle 12 has already been mentioned. Notably, it removes the linkage, present in both Stockholm and in the WCED report, with goals of the NIEO to restructure the international economic system. It also equates sustainable development with economic growth when it states that an open international economic system "would lead to economic growth and sustainable development in all countries" and would therefore "better address the problems of environmental degradation" than, presumably, a less open international economic system. This goes much further than the Brundtland Commission, which, although it supported economic growth, especially in developing countries, saw environmental protection as a necessary condition of sustainable development. Nonetheless, the Rio Declaration is not as large a step from Brundtland as some analysts suggest. It merely legitimated one particular interpretation or pathway in the operationalization of sustainable development already discernable in the language WCED used. Pallemaerts (1996:633) is correct that the Rio Declaration "confers on economic growth a new ecological legitimacy" but overstates the case that such legitimacy was not already implied in the WCED report.

The one norm that implies that any obligations toward the environment might operate in anything but a liberal market context is Principle 7, which recognizes the "common but differentiated responsibility" of developed and developing states toward the pursuit of sustainable development. While it does not contradict liberal environmentalism, it does harken back to NIEO goals of differential obligations of the North and South and hence some possible interference in what might be the most economically efficient means of dealing with global environmental problems. This principle can also be found in articles 3(1) and 4(1) of the FCCC and is a fundamental element of the implementation of the treaty, which creates different obligations for developed and developing states.

The main operative provisions of the FCCC deserve mention in this regard since the operationalization of "common but differentiated responsibility" still appears to fit with using or creating markets and liberal economic norms more generally. Article 4(2)(a and b) spell out commitments. In line with common but differentiated responsibilities, Article 4(2)(a) obligates developed states to "tak[e] the lead" in modifying their greenhouse gas emissions, but to do so while recognizing, *inter alia*, "the need to maintain strong and sustainable economic growth." It further states that, "Parties may implement such policies and measures jointly with other Parties." This idea of "joint implementation" was shown earlier to fit with the marketization of environmental protection. Hence, even the commitment in article 4(2)(b) by developed countries to "aim" to return to 1990 emission levels of carbon dioxide and other greenhouse gases not controlled by the Montreal Protocol by 2000 can be achieved "individually or jointly."

In terms of how states should manage national and international environmental problems, Principle 11, in combination with the Precautionary and Polluter Pays Principles (15 and 16), makes clear that any such management must fit into a general program that promotes economic growth and liberal markets. Principle 11, for example, says states "shall enact effective environmental legislation" but that "standards, management objectives and priorities should reflect the environmental and developmental context to which they apply." Similarly, the PPP, in order to avoid any misunderstanding, must only be applied "without distorting international trade and investment." That meaning fits precisely with the way the principle developed within the OECD.

The PPP is reinforced by chapter 8 of Agenda 21, on integrating environment and development in decisionmaking. It proposes that a legal framework for sustainable development should "not only [act] through 'command-and-control' methods, but also [act] as a normative framework for economic planning and market instruments" (Agenda 21:8.13 in IDRC 1993). Chapter 8 also explicitly promotes more widespread use of market mechanisms and measures aimed to internalize environmental costs, both of which follow from the PPP. Although such measures are to be "complementary" to regulatory approaches, a quarter of chapter 8 is devoted to market instruments and the overall normative thrust is to "include, wherever appropriate, the use of market principles in the framing of economic instruments and policies to pursue sustainable development" (Agenda 21:8.31 [c]). Since 1992, the UN Commission on Sustainable Development has reiterated PPP on several occasions and the discussion earlier has shown that it, and concepts associated with it, form the normative basis of a wide range of environmental policies and programs. Grubb et al. (1993:113) sum up the importance of chapter 8 to future environmental policy as follows:

> In setting out the general measures and form of policies which need to be adopted in pursuit of sustainable development, this little-publicized chapter—finalized at PrepCom IV—forms potentially one of the most powerful of all individual chapters in Agenda 21. . . . It reflects a strong move towards consideration of economic instruments for environmental policy, and comes very close to a global endorsement of a "polluter pays principle."

Similarly, Principle 15 for the first time endorses the Precautionary Principle in a global declaration, although the approach is qualified in two ways. First, it shall be applied "by states according to their capabilities." Second, a "lack of full scientific certainty shall not be used as a reason for postponing *cost-effective* measures to prevent environmental degradation" (emphasis added). The endorsement of "cost-effective" measures implies that the PPP and precautionary approach should be applied together. The principle did not originate at Rio, but has roots in German environmental thought and had previously appeared in embryonic form in regional documents and declarations prior to Rio, dating back to at least the Ministerial Declaration at the second International North Sea Conference in 1987 (Birnie 1992:88; O'Riordan and Cameron 1994). Since its legitimation at

Rio, however, it has gained some prominence in international law and discourse, appearing in a wide range of conventions including those on climate change and biodiversity, a number of international agreements on fish stocks, the 1996 Protocol to the London Dumping Convention, the Biosafety Protocol of the Convention on Biological Diversity, and in the fifth environmental action programme of the European Union.[50]

Not all norms listed in the Rio Declaration fit neatly into the norm-complex of liberal environmentalism. For example, a number of principles address the need to increase participation in environment and development decision making and access to information (Principle 10) and to encourage participation of various societal groups including women (Principle 20), youth (Principle 21), and indigenous people (Principle 22). These principles are consistent with a wide range of multilateral activities and promotion of these norms in other contexts at the international and domestic levels. These norms are not central to (and do not contradict) the main argument put forward here, but should be acknowledged as potentially important components of the future development of sustainable development thinking.

In addition, purely political principles found their way into the Declaration, such as the nod to the plight of the Palestinians in Principle 23, which called for the protection of the environment for "people under oppression, domination and occupation." Similarly, the calls for international cooperation (Principle 27), protection of the environment during warfare (Principle 26), and the indivisibility of peace, development and environment (Principle 25) represent general aspirations rather than specific goals.

I describe the resulting norm-complex of liberal environmentalism below to facilitate comparison to the norm-complex from Stockholm. Principles in brackets refer to the Rio Declaration. The summary below is followed by Table 1 on the evolution of norms from 1972–1992.

State Sovereignty and Responsibility

1. The basic principles (2, 13, 14) remain unchanged from Stockholm with two important exceptions. First, the obligation to notify others of potential environmental harm not accepted at Stockholm is entrenched (Principles 18 and 19). Second, added to a state's right to exploit its own resources pursuant to environmental policies is to do so pursuant to development policies, which arguably upsets the balance

struck at Stockholm between sovereign rights to exploit resources and environmental protection (Pallemaerts 1994:5).

Political Economy of Environment and Development

2. The new equity principle of "common but differentiated responsibilities" for developing and developed countries replaces Stockholm's emphasis on the latter and Brundtland's on the former (Principle 7). Two imperatives follow:
   a. The right to development is entrenched which is generally consistent with Brundtland's imperative to revive global growth (Principles 3–5). Growth and development have precedence over environmental protection if the social and economic costs are too high for developing countries (Principle 11).
   b. "Unsustainable" patterns of production and consumption should be reduced and eliminated (Principle 8).
3. Free trade and liberal markets are supported unequivocally with no reference to interventions such as commodity price stabilization. Free trade and environmental protection are presumed to be compatible (Principle 12).
4. Technology transfer is essentially left to market mechanisms, except for the least developed countries (Principle 9) (see Haas, Levy and Parson 1992:28–32).
5. The same environmental cooperation ethic remains (conserve and enhance resources for present and future generations), but human beings should be at the center of such concerns (Principles 1, 7, and 27).

Environmental Management

6. Although environmental impact assessments are endorsed (Principle 17) the primary management norms are the PPP (Principle 16) and Precautionary Principle (Principle 15). The former promotes a preference for market-based instruments over purely regulatory methods. As argued, the Precautionary Principle is fully compatible with this approach. The theme of cost-effectiveness also runs through the range of UNCED documents and the statement on integrating environment and development in decision making (Agenda 21, chapter 8) supports this trend, suggesting that environmental assessments are to be considered on cost/benefit criteria and with PPP in mind.

TABLE 1 The Evolution of International Environmental Norms: 1972–1992

	STOCKHOLM 1972	WCED 1987	UNCED 1992
STATE SOVEREIGNTY AND LIABILITY	1. Sovereignty over resources and environmental protection within state borders. Responsibility for pollution beyond state borders. (Principles 21–23).	1. Unchanged.	1. Unchanged (Principles 2,13, and 14) except: a) advanced notification of potential environmental harm (Principles 18 and 19); b) state right to exploit resources is to be pursuant to *development* in addition to environment policies.
POLITICAL ECONOMY OF ENVIRONMENT AND DEVELOPMENT	2. Developed and developing countries differ on sources of and solutions to environmental problems. (Principles 11–13).	2. States have the following *common* responsibilities: a) revive global growth b) participate in shared responses to global environmental problems.	2. Common but differentiated responsibility of developed and developing countries. (Principles 3, 7, and 11). Development takes precedence if costs of environmental protection too high (Principle 11).
	3. Balance free trade with commodity price stability. (Principle 10).	3. Free trade plus an emphasis on global growth balanced with managed interventions and commodity price stability.	3. Free trade and liberal markets. Environment and free markets compatible. (Principle 12).
	4. Environmental protection requires substantial transfers of technology and resources to developing countries. (Principles 9 and 20).	4. Unchanged plus specific proposals such as a tax on use of the global commons.	4. Transfers left primarily to market mechanisms, except for least developed countries.
	5. States should cooperate to conserve and enhance global resource base. (Principles 1–7 and 24).	5. Multilateral cooperation for global economic growth as necessary for other goals.	5. Same as WCED plus human centered development. (Principles 1, 7, and 27).
ENVIRONMENTAL MANAGEMENT	6. Command-and-control methods of regulation favored over market allocation in national and international planning. (Principles 13 and 14).	6. Mix of command-and-control and market mechanisms. Polluter Pays Principle (PPP) endorsed.	6. Market mechanisms favored. PPP and Precautionary Principle. (Principles 16 and 15).
NORM-COMPLEX	**ENVIRONMENTAL PROTECTION**	**MANAGED SUSTAINABLE GROWTH**	**LIBERAL ENVIRONMENTALISM**

LIBERAL ENVIRONMENTALISM AFTER RIO

The aftermath of Rio was a disappointment for most environmentalists. Whereas the first oil shock, recession, and debt took their toll following Stockholm, this time around a new set of global crises pushed the environment lower on foreign policy agendas. In the wake of the Cold War, the proliferation of civil and ethnic conflict, problems of post-communist transition, and financial crises in Mexico and later Asia diverted the resources and commitment needed to follow through on promises made in 1992. In this section I show that, despite the poor record of environmental achievements following Rio, liberal environmentalism remained the dominant governing norm-complex for global environmental concerns, and its institutionalization actually increased.

The main question I address is whether or not practices and actions actually taken reinforced liberal environmentalism. I find that most did, with a few qualifications. For example, actual practices sometimes fit with alternative norms as much as with liberal environmentalism, and lip service did not always translate into action on the ground. In addition, negotiations since Rio on a number of environmental issues revealed that some European states remained uncomfortable with policies that seemed to place too much faith in the compatibility of the market and environmental protection. Meanwhile, developing countries continued to pursue traditional goals of technology transfer, aid, and differential obligations alongside of support for elements of liberal environmentalism. Despite these areas of contestation, however, liberal environmentalism continues to prevail as the dominant norm-complex because few states show a willingness to reopen the normative consensus agreed to at Rio, and these norms continue to guide specific policies, research, and action.

Below, I focus on two major events chosen because they epitomize the normative direction of environmental governance since Rio and because of their importance to global environmental politics more generally. They also reflect policy directions in the most recent wave of international environmental action, and cover issues on which states and international organizations have devoted much high-level attention resources. First, I will discuss the UNGA's June 1997 Special Session to Review Implementation of Agenda 21 (UNGASS) in New York, the political review and assessment five years later of post-Rio achievements and failures. Although much lower-profile than the Earth Summit, it was the first high-level meeting following Rio to cover the entire range of global environment and development issues raised in Agenda 21. Participants included 53 heads of state and government, as well as ministerial and other high-level representation. Second, I will discuss international action to address climate change. Both demonstrate that liberal environmentalism remains a powerful normative underpinning of international environmental governance, even if its success in producing environmental action can be questioned.

The 1997 UN Special Session: The Earth Summit + 5

Two facets of the special session stand out. First, as summed up in one analysis: "The 'Earth Summit + 5' proved to be a sobering reminder that little progress has been made over the past five years in implementing key components of Agenda 21 and moving toward sustainable development" (IISD 1997a). Whereas a number of successes could be identified—the creation of more than 100 national sustainable development bodies, the initiation of 1,800 local Agenda 21s worldwide, and the entry into force of several treaties (Brown 1998)—political will and financial commitment since Rio appeared lacking. Second, despite this assessment, the special session reinforced the norms of liberal environmentalism as the appropriate guide to address global environmental concerns. I address each facet in turn.

From the perspective of developing countries, the most glaring lack of commitment since Rio concerned the areas of finance, technology transfer, technical assistance, and capacity-building. Many states singled out the sizeable expansion of private financial flows as the major change in the international political economy since UNCED that could explain these difficulties. Whereas delegates noted that this change produced greater investment in a limited number of developing countries, the debt situation remained a major constraint to achieving sustainable development in many others. Meanwhile, the technology gap between developed countries and, in particular, the least developed countries had widened (IISD 1997a).

The special session also identified lack of progress on a number of specific environmental issues. However, the protection and sustainable use of the world's forests stands out as an example, especially given the high hopes for a breakthrough on the issue generated prior to the UNGASS. Perhaps apart from desertification, where a convention was successfully negotiated in 1994, the forest issue received the greatest attention of the substantial concerns left unresolved in 1992. Indeed, many expected a concrete agreement on forests to be the showcase achievement of the special session. That this did not occur, and the reasons for the lack of progress, reveals both the disappointment of UNGASS and the continued effects of liberal environmentalism.

Although a number of initiatives on forests made some headway following 1992,[51] the only one to focus serious attention on renewing attempts to build consensus on a global convention was the establishment of the Intergovernmental Panel on Forests (IPF) in 1995. After two years of intensive intergovernmental discussions sponsored by the UN Commission on Sustainable Development—which were to wrap up in time to forward recommendations to UNGASS—states at the special session decided only to continue the intergovernmental policy dialogue on forests through the *ad hoc* open-ended Intergovernmental Forum on Forests (IFF). Like the IPF, the IFF was directed to "identify the possible elements of work toward consensus on international arrangements and mechanisms, for example, a legally-binding instrument," and report to the Commission on Sustainable Development at its eighth session in 2000.

No one originally envisaged the IPF as primarily a vehicle to build consensus for a convention. However, momentum grew as its deliberations got underway, perhaps because a convention would have been a tangible outcome from the long, complex process to which states had committed much time and resources. Although discussions advanced on a number of technical issues, divisions present in UNCED negotiations resurfaced immediately at IPF-1 in September 1995.[52] These tensions revolved essentially around any issue that conflicted with the liberal environmentalism compromise. Most noticeably, the G-77/China bloc resisted any proposal they viewed as potentially leading to the loss of national control of forests or forest products. The issue of sovereignty arose most forcefully in discussions over trade and environment and the related issue of certification and labeling schemes—which promoted sustainability by certifying forest products as meeting agreed to standards. Developing countries strongly opposed any mandatory schemes, viewing them as primarily aimed at tropical forests. With some support from Northern producer countries, developing countries argued that any such scheme must only occur as part of a broader framework that promoted market access and freer trade in forest products, including those from the tropics. Strong opposition also arose over any proposals or discussions around unilateral bans and restrictive measures, with strong support for norms of free trade from most countries in North and South. Underlying all these discussions was the tension between North and South over the inclusion of all forests in discussions rather than targeting tropical forests only.

By IPF-4, when explicit discussions related to a convention finally occurred, the lack of progress led to the disintegration of the pro-convention coalition of states and NGOs. Opposition to a convention arose from an unusual coalition of the United States, Australia, Brazil, some other developing countries, and several environmental groups, including the Worldwide Fund for Nature, Greenpeace, and Friends of the Earth.[53] Environmental NGOs feared that in the six to ten years it might take to negotiate and ratify a convention, governments would only relax their protection of forests. They argued that the depth of disagreement over the shape of a convention would produce a lowest common denominator agreement, amounting to little more than "a loggers' charter" that would emphasize only the economic value of timber in commodity markets. Thus their goal of an agreement that linked forests to broader concerns such as biodiversity and climate change appeared out of reach. Environmental groups argued that a better approach would be to build on voluntary initiatives and current agreements, which they believed already provided a legal basis for sustainable forestry and protection.[54]

Not much changed on progress toward increased international cooperation on forests after states failed to reach agreement on even the basis of a convention at UNGASS. After three years of IFF discussions that included a mandate to build consensus toward a legally binding instrument, and despite the IFF-sponsored Canada–Costa Rica initiative that aimed specifically to build consensus on elements of a global forest convention, the IFF was no closer to consensus on a convention than its predecessor when it reported to the Commission on Sustainable Development at its eighth session in April–May 2000. Although progress could be identified on the implementation of some recommended actions from the earlier IPF process, the core issues of disagreement that frustrated agreement at UNCED remained largely unresolved. Notably, decisions on how to proceed on trade- and sovereignty-related issues reinforced liberal environmental norms, which, in this case, militated *against* an agreement that would satisfy environmental concerns of sustainable forest management (SFM). The difficulty of reconciling these values in practice remained a major obstacle to a convention. The IFF simply concluded at its final meeting in February 2000 (IFF-4) that "trade measures intended to promote SFM should not constitute arbitrary or unjustifiable discrimination or a disguised restriction on trade" (IISD 2000b). Voluntary certification and labeling schemes would be considered (discussed further below), but only if they could be shown not to unjustifiably limit market access and that work be done to ensure adequate transparency and nondiscrimination in their design and operation, thus supporting the overarching goal of trade liberalization and World Trade Organization (WTO) norms. The IFF also reached consensus to support full-cost internalization of forest products and services, reinforcing the polluter or user-pay principle (IISD 2000b).

The end result was a decision by states at the Commission on Sustainable Development to more-or-less institutionalize the IFF. They proposed the creation of a UN Forum on Forests (UNFF) within the UN system that would carry on work to implement existing agreements and initiatives from the IPF/IFF process. It would also address financial resources for implementation, and would again "consider," within five years, based on an agreed to assessment of the arrangement, "recommending parameters of a mandate for developing a legal framework on all types of forests." Such language suggests a less than sanguine prospect for the development of such an instrument any time soon. Finally, again reinforcing key deep norms of liberal environmentalism, states also agreed the UNFF would be intergovernmental, and referred to Agenda 21 and the Rio Declaration for its principled basis (IISD 2000c). The UNFF became a subsidiary body of ECOSOC in October 2000.

The poor prospects for a convention, I would argue, stem from the difficulty of managing a resource such as forests that requires action that conflicts with norms agreed to at Rio. For example, because forests are perceived as a global commons problem, but fall generally within particular states, actions required appear to threaten sovereignty over resources, free trade, and other norms of liberal environmentalism. Thus, building a legitimate basis for action within the current norm-complex has proven difficult. Still, no alternative has been put forward.

This observation points to the second facet of UNGASS: that despite the lack of progress on a number of issues, the special session reinforced the normative consensus institutionalized at Rio. The Programme for Further Implementation of Agenda 21 negotiated at UNGASS reaffirms states' commitment to Agenda 21 and all the principles in the Rio Declaration on Environment and Development (UNGA 1997; ENS 1997). Delegates attributed blame for lack of action not to flaws in Agenda 21 or guiding norms, but to external events that put new pressures on states and resources that otherwise might have been devoted to fulfilling Rio's promises. Although many delegates lamented this state of affairs, they did not fundamentally challenge the view that the Rio norms could marshal such forces if only enough political will could be generated.

For example, many states noted that the major increase in the globalization of economic activity decreased the significance of traditional means of promoting development, such as Official Development Assistance, which had declined since 1992. At the same time, the UNGASS program of action noted the opportunities that private investment can provide if it could be channeled toward the goals laid out in Agenda 21 (UNGA 1997). While the program of action recognized that implementing policies in areas such as trade and the environment, corporate responsibility, and incentives for environmentally-friendly investment faced a variety of hurdles, it also affirmed that the norms institutionalized at Rio fit with this new reality.

Take the case of the Polluter Pays Principle. It continues to be seen by industrial countries as an important component in attempts to link free trade and environmental concerns and UNGASS negotiations reinforced its primary implication of eliminating subsidies and internalizing environmental costs. For example, the Programme of Further Implementation includes internalizing environmental costs and reducing or eliminating subsidies as primary means both of addressing unsustainable production and consumption patterns and of implementation of Agenda 21 as a whole.[55] In addition, it identifies the reduction or reformulation of subsidies as a primary means of implementing recommendations on a number of specific issues ranging from fisheries management to energy production and consumption.

Whereas most states fought against reopening discussions on guiding norms, practical issues of implementation both inside and outside of UNGASS have and continue to foster debates, which may make putting the norms into practice difficult. Contestation continues to occur particularly over the meanings of norms. Staying with the example of polluter pays, some developing countries continue to stress PPP's implications for equity and liability. They interpret PPP to require developed countries to shoulder greater responsibility, and costs, for environmental action given their greater historical contribution to environmental degradation. Some states have used this argument in climate change negotiations, for example. This formulation links PPP to the norm of "common but differentiated responsibility" of developed and developing countries to protect the global environment. A second challenge comes from PPP's interpretation as a liability norm that would require compensation for accidental or "residual" pollution (Pearson 1994). Nonetheless, while these interpretations may pose difficulties for its implementation in specific agreements, such as the GATT/WTO, they have not undermined the broad legitimacy of the principle—or a norm with similar propositional content—as a basis for linking trade, economic activity, and environmental concern.

Similarly, discussions on trade and environment, while far from resolving a variety of outstanding debates, rarely stray from liberal environmental norms. Perhaps unsurprisingly, research and discussions within both the OECD and WTO, which dominate international attempts to address the issue, reinforce liberal environmentalism even though their studies tend to admit that actual trade practices do not live up to the ideal. Reports by the OECD Joint Session of Trade and Environment Experts (May 1995) and the Committee on Trade and the Environment to the WTO (November 1996) reached similar conclusions that, "in general terms, trade liberalization will have a positive impact on the environment by improving the efficient allocation of resources, promoting economic growth, and increasing general welfare, provided effective environmental policies are implemented."[56] Both reports also refer explicitly to Rio Declaration Principle 12—on the compatibility of free trade and environmental protection and against unilateral environmental measures—as legitimating this position. The WTO report does suggest some room for measures aimed specifically at products covered by multilateral environmental treaties. In contrast, the OECD report comes out strongly against any trade measures, arguing they "are very rarely the primary or first best instrument for achieving environmental objectives" (quoted in Reiterer 1997:74).

The most likely way around such rules is through measures that address Processing and Production Methods (PPMs), which can potentially operate within free trade systems, but which remain controversial and are mentioned but not explicitly covered under the Technical Barriers to Trade (TBT) Agreement of the WTO.[57] The recent cautious support of voluntary certification and labeling schemes in various fora as a means to promote sustainable management of forests offers a good example of how PPMs might work in practice. (Eco-labelling is allowed under certain circumstances and covered under the TBT Agreement, but the rules still require further clarification and are subject to ongoing discussions in the WTO Committee on Trade and Environment [CTE]).[58] The Forest Stewardship Council (FSC) sponsors one of the best-known voluntary certification and labeling programs in the forestry issue area. Spearheaded by the Worldwide Fund for Nature (WWF), FSC accredits organizations (certifiers) who must perform evaluations to see if a company's forest operation matches ten established principles and criteria for "well-managed" forests. More specific regional standards are then developed based on these broader principles. By 1998, FSC had accredited certification companies based in the United Kingdom, the Netherlands and the United States. In addition, WWF has successfully organized the creation of buyer groups for certified wood products in the UK, Holland, Belgium, Austria, and the United States (Hansen 1998).

Recent attention to certification and labeling schemes can be understood in the context of their fit with liberal environmentalism, as they aim essentially to internalize environmental costs by including them in the cost of products certified. Since consumers, if educated by the eco-labeling processes, would presumably favor such products, the market would provide economic incentives to live up to the labeling criteria. The market for forest products would then operate with the "right" prices. A number of problems have yet to be resolved with certification, and certified forests still reflect a very small percentage of the world's total. Yet, the attempt to do an end-run around sovereignty and to avoid restrictive trade measures by going directly to the marketplace provides a good example of the direction liberal environmentalism is likely to push, given the way this norm-complex appears to constrain state action on this issue.[59] It remains to be seen whether such schemes that operate with the market can sufficiently achieve sustainability or other environmental values, including linkages to broader ecosystem concerns such as biodiversity. Moreover, since achieving these broader goals may move further into the area of non-product-related production and processing methods, they may be more difficult to achieve under current WTO rules, even if they are fully consistent with cost internalization.

The rise in what Jennifer Clapp calls the "privatization of global environmental governance" reflects another recent trend in environmental practice that attempts to work within the norms of liberal environmentalism (Clapp 1998). By this she means the growing number of voluntary codes of conduct and private or hybrid (mixes of state and non-state member) standards setting bodies—such as the International Organization for Standardization's ISO 14,000 standards—that address environmental concerns. While NGOs have so far managed to take the lead in the case of forestry, most of the processes are industry-led in an attempt to avoid state-determined environmental regulation. Standards such as those set by the ISO are recognized by the WTO as legitimate public standards. Like certification schemes, such processes can effectively work within liberal international norms that call for a reduced role for the state (Clapp 1998:298).

The wide recognition of such standards by firms, states, and international organizations suggests institutionalization of liberal environmentalism continues to increase. Such private authorities also reveal an important contradiction in the liberal environmentalism compromise at Rio—that sovereign control over environmental policies might not be fully compatible with liberal norms. Thus, the proliferation of such authorities has the potential to be a site of contestation. Nonetheless, since such authorities usually avoid any attempt to impose policy directly on states, they can effectively divert pressure from states to implement unpopular policies and pose little threat to sovereign authority in practice.

The Kyoto Protocol

Perhaps no better example of the effects of liberal environmentalism exists than the signing of the 1997 Kyoto Protocol to the Framework Convention on Climate Change (FCCC).[60] Although the Montreal Protocol on Substances that Deplete the Ozone Layer also permits international trading of CFC quotas,[61] the Kyoto Protocol is the most ambitious attempt to date to implement market and other economic mechanisms at the global level that I have identified as a key component of liberal environmentalism. Given that climate change has generated more high-level political concern than any other global environmental problem, greater devotion of resources toward research, and now promises of action likely to touch on a wide range of economic and environmental activities, the framing of the problem and solutions in liberal environmental terms is particularly significant.

The compromise behind the Protocol links quantitative reductions or limits in greenhouse gases in developed countries[62] to three main market mechanisms that involve transferring "credits" for emissions to help countries meet their targets: emission trading among developed countries; joint implementation (JI) among developed countries, where emission reductions financed by foreign investments would be credited to the source country; and a Clean Development Mechanism (CDM) to finance projects in developing countries, where the investor, from a developed country, would receive "certified emissions credits" for emission reductions produced by the project in the developing country.[63] The Kyoto mechanisms all work on the same basic principle: that assigning property rights to emissions and creating a market that allows them to be transferred will enable emission reductions to be achieved where it is most efficient, or cheapest, to do so. The impact on the atmosphere should be the same regardless of where cuts are made. They can be considered "market" or "incentive-based" mechanisms because they rely on the establishment of a market for emission credits to create price signals, and thus incentives, for buyers, sellers, and investors, as long as abatement costs vary across countries. Although the extent to which the mechanisms can be used to reach a country's target and technical details of the mechanisms were two of the issues states failed to reach agreement on at the sixth Conference of the Parties (COP-6) in the Hague (November 13–25, 2000),[64] and countries such as United States will not even consider ratification until these issues are resolved, the basic shape of environmental governance Kyoto endorses is unlikely to change.[65]

Indeed, the United States was largely behind the breakthrough idea to link binding targets to market mechanisms that led deveoped countries to commit to an average 5.2 percent reduction in GHG emissions from 1990 levels by 2008–2012. This position started to unfold in 1996, as indicated by U.S. Under Secretary of State Timothy Wirth's speech to the Second Conference of the Parties in July 1996, where he made the link explicit:

Based on these principles—encompassing environmental protection, realism and achievability, economic prosperity, flexibility, fairness and comprehensiveness—the United States recommends that future negotiations focus on an agreement that sets a realistic, verifiable and binding medium-term emissions target. We believe that the medium-term target must be met through maximum flexibility in the selection of implementation measures, including the use of measures such as reliable activities implemented jointly and trading mechanisms around the world (Wirth 1996 and author's interview).

As noted earlier, the Clinton White House strongly supported research and policies along these lines, and Wirth himself advocated this approach. Despite initial concerns expressed about this linkage from a number of states, the strong U.S. stance and leadership on this issue and the movement of many states to accept the linkage since 1997 makes it likely that the emphasis will continue to be on market-friendly mechanisms and liberal environmentalism as an underlying normative framework for the FCCC. Indeed, a variety of major global corporations and international organizations, including institutions such as UNCTAD where developing country concerns dominate, are vying to position themselves to take advantage of these mechanisms.[66]

Notably, even proponents of carbon "sinks" or sequestration frame them as resting on the same normative logic as the Kyoto mechanisms, that is, on efficiency and cost-effectiveness as important criteria in achieving environmental goals. For many countries, this same logic translates into domestic implementations regimes that emphasize flexibility and incentives rather than regulations. Incentives may range from tax incentives, to research funding, to government procurement to encourage technological innovation in the energy sector, to formal incentive mechanisms such as internal trading schemes or schemes to give companies credit for early action. In addition, this logic supports voluntary or self-regulatory initiatives among industry, and an overriding sensitivity to the international competitive implications of any domestic implementation policies.

CONCLUSIONS

This chapter and the last have traced through norms of international environmental governance and demonstrated their evolution toward the compromise of liberal environmentalism. The United Nations Conference on Environment and Development legitimated this norm-complex which now dominates practices of many states, international organizations, and cooperative arrangements that govern responses to global environmental problems.

Admittedly, not all analysts examining the content of international governance would have focused on the political economy of environment and development as I have. Others have focused on the push toward local participation in environmental decision making, environmental security, or simply the increased scope of international cooperation and activity around environmental problems that cross borders. Although these and other changes in the nature of environmental governance are important, I have argued that the major thrust of the institutionalization of "sustainable development" has been toward liberal environmentalism and that this set of norms encapsulates the main ideas that UNCED legitimated. To focus on other factors misses the core of the compromise at the heart of the norm-complex institutionalized at Rio.

Furthermore, I have argued that liberal environmentalism marks a significant shift from earlier attempts to address global environmental problems and to link environment and development. The norms and practices that followed from the Stockholm conference tended merely to juxtapose environment and development and often implied a suspicion, if not outright hostility, toward market forces. Solutions to environmental problems were most often framed in terms of the need for regulation and intervention, as were solutions to adapting development strategies to address environmental problems. The Brundtland Commission attempted a new synthesis of environment and development that put economic growth at the center of strategies for sustainable development. It proposed a mix of market forces, redistributive policies, and environmental interventions to promote growth of a sustainable kind, in what I have called a norm-complex of managed sustainable growth.

The Earth Summit institutionalized one major pathway from Brundtland. It entrenched the idea that market forces can be compatible with environmental protection and that a liberal economic order is best suited to achieving environment and development goals. It showed suspicion toward the types of global management implied by the Brundtland Commission, or by norms such as Common Heritage. Instead, the Earth Summit agreements reflected a faith in the market, or in the increased adaptation of human activity to market norms, as the preferred means to solve environmental problems.

Indeed, one of the major omissions from Agenda 21 was the regulation of multinational corporations, which might have restricted freedom in the global marketplace. That omission occurred in the context of the active participation of multinational corporations in the conference and the close relationship of at least one powerful industry lobby—the Business Council for Sustainable Development (BCSD)—with the conference secretariat.[67] Industry played a dual role at UNCED, with some groups lobbying hard to prevent any regulation that might threaten their short-term interests, while other groups, including the BCSD, argued that industry could play a positive role via self-regulation.[68] In the end, industry was enlisted to voluntarily engage in good practices and their freedom of activity was apparently seen as important for the overall goals of the liberal economic order.

Thus the Earth Summit outcomes emphasized norms consistent with free trade, the Polluter Pays Principle, and Precautionary Principle, and promoted market mechanisms to address environmental problems. Furthermore, UNCED reinforced state sovereignty and control over global resources and placed human beings squarely at the center of global environmental governance. The next two chapters offer possible explanations of why the ideas associated with liberal environmentalism prevailed.

Chapter 4
EPISTEMIC COMMUNITIES, SCIENCE, AND
INTERNATIONAL ENVIRONMENTAL GOVERNANCE

SCIENTISTS AND THEIR FINDINGS mattered in the development of international environmentalism. The very nature of global environmental problems as uncertain and complex ensures that technical expertise is called upon for understanding and advice, and that scientific discovery can bring previously unknown problems to the attention of policymakers. It would be truly remarkable if scientists played no role—akin to suggesting that agricultural experts played no role in world agricultural or food programs or that medical doctors or researchers played no role in world health programs. If the research question of interest was simply "did scientific and technical knowledge on specific environmental problems influence international cooperation on those problems," or "did the growth in scientific knowledge or a rationalized scientific culture play some role in the increase in international efforts to address environmental concerns," the answer would surely be yes, or at least a qualified yes given that other factors also played a role.[1] But the research question that guides this study is not on the rise of environmentalism generally, nor is it mainly to explain cooperation on a specific environmental problem. Rather, the question is why did the *content* of global environmental policy, or the *appropriate* way to understand, address, or manage global environmental problems, evolve as it did? In other words, why did some ideas prevail over others to guide global environmental governance toward liberal environmentalism?

I begin with a focus on scientists, or expert groups more generally, because so much of the scholarship on global environmental problems either assumes or sets out to show the central importance of scientific communities and knowledge to environmental governance. This scholarship also contains an implicit prescriptive element. Whereas very few scholars naïvely argue, as Underdal (2000a:5) puts it, that "science is seen as carrying the torch of light, guiding what Plato referred to as 'philosopher kings' in their altruistic search for the common good," many authors who view rational science and expertise as necessary for understanding nature, and for the effective management of international environmental problems, do seem to pin their hope for improved environmental governance on the progressive influence that scientific communities can exert to modify state interests.[2] The question of whether this hope is well founded, as well as the rationalist basis of this hope, both deserve to be critically assessed.

As a lens through which to address these questions, this chapter tests in detail a prominent explanation for the evolution of environmental governance that focuses on the influence of expert or "epistemic" communities (Haas 1989, 1990, 1992a, 1992b, 1996; Adler and Haas 1992). These communities are bound together in a common policy enterprise and empowered by shared causal and principled beliefs. They need not be composed strictly of natural scientists, but their legitimacy, and thus their power to affect governance, must stem from shared notions of validity, or shared criteria for weighing and validating knowledge in the domain of their expertise (Haas 1992c:3). The assumption is that the commitment to professional methods and norms of scientific inquiry gives policymakers and the public confidence in the autonomy and integrity of scientists (Underdal 2000a:10). Thus, this explanation argues that scientific communities can influence and shape international policies because policymakers and governments recognize epistemic communities as legitimate sources of knowledge and expertise, and thus call on them to achieve policy goals in issues characterized by uncertainty and complexity. The global environment is clearly such an issue.

I test this hypothesis in particular for the three reasons listed in the introduction: its clear explanatory framework of how scientific knowledge translates into changed patterns of state behavior and international inter-actions; its influence on literature on the role of ideas in international envi-ronmental coordination and international relations more generally;[3] and because environmental governance should be a most-likely case, indeed, the paradigmatic case for the hypothesis. Thus, its failure in this case would provide strong evidence against the general validity of the hypothesis, since the expectation would be for it to also fail in less hospitable circumstances.

The epistemic communities literature also offers a good entry point into a broader discussion about the role of science and scientists in environ-mental governance. While the chapter specifically sets out to rigorously test a specific hypothesis of importance to political scientists, the insights this test generates will be of interest to anyone interested in the possibilities and limits of the influence of scientific knowledge on global environmental policymaking and international affairs more broadly. As will be seen, the findings of this chapter strongly contradict the conventional wisdom. They demonstrate that the central place of scientists in constructing action on issues that require technical and scientific expertise has been greatly exag-gerated. Moreover, contrary to expectations in much of the mainstream scholarship, I find that scientific research and the uses to which it is put are as strongly shaped by existing social structures as vice-versa.

Given these shortcomings, the next chapter examines the role of eco-nomic ideas and puts forward an alternative explanation that better ac-counts for the interaction of ideas and social structures in global gover-nance. It focuses on social structural pressures that favor the selection of some ideas over others in the institutionalization of international norms. In many ways, the next chapter can be read as an attempt to recapture the core insight from the broader agenda of the epistemic communities litera-ture, that agency, legitimacy of ideas, and broader "epistemes" (dominant ways of looking at social reality or a set of shared symbols and references) can be important sources of continuity and change in international politics and can even (re)shape understandings of state interests.[4] However, the ap-plication of the epistemic communities literature, as it evolved to focus mainly on agency and the problem of cooperation or policy coordination, has suffered because of its inattention to the constraints of the internation-al system. Recasting some of its insights in a broader theory of the interac-tion of ideas and international social structure offers a fuller understand-ing of the opportunities and constraints that new ideas face.

The separation of scientific and economic ideas may at first appear arbi-trary or artificial. For example, economists and natural scientists some-times interact in the policy process and even in the technical work on ecological problems. Also, individual scientists and economists might be influenced by ecological ideas and both groups might be considered "epistemic communities" in their own right. However, treating both equally as epistemic communities undermines the logical basis of the explanation—that a single community is granted legitimacy based on its claim to authoritative and policy-relevant knowledge in a certain issue area. If more than one such community exists, the reason for adopting the position advocated by such a group could not be accounted for simply by looking at its privileged position owing to its knowledge claims. Thus, in this chapter, I focus on the strongest group identified in the literature—scientists, and especially those scientists loosely considered sympathetic to "scientific ecology."

The chapter begins with a discussion of the precise claims of the epistemic communities literature in order to clarify how evidence should be evaluated. The remainder of the chapter traces the influence of scientists, determines whether they constituted an epistemic community, and assesses the influence of such a group in the lead-up to and deliberations during the three key turning points in environmental governance in 1972, 1987 and 1992. By focusing on what scientists actually did and the effects of those actions, I also hope to illuminate the interaction of science and global environmental governance rather than merely offer a critique of the epistemic communities hypothesis.

EXPLAINING NORM CREATION AND CHANGE WITH EPISTEMIC COMMUNITIES

Stated formally, an epistemic communities explanation asserts *that scientific consensus within an epistemic community, "politically empowered through its claims to exercise authoritative knowledge and motivated by shared causal and principled beliefs,"*(Haas 1992a:41) *and its promotion of norms derived from that consensual knowledge, leads to the adoption of its ideas over others as guides to appropriate behavior.* In this context, the relevant question is, did ideas promoted by key epistemic communities prevail over other ideas? Or, more generally, to what extent do expert groups determine the observed content of governing norms or specific policies those norms support?

The clearest substantive application of the epistemic communities hypothesis to explain international environmental governance comes from Peter Haas, who argues that an epistemic community formed around a "scientific ecology" research program (1989, 1990, 1992a, 1996). Scientific ecology is "distinguished by its systems perspective on environmental, social and economic problems; reflecting a multi-sectoral approach and a normative commitment to environmental preservation." Furthermore, members of this epistemic community have "sought to develop social laws from their understanding of the laws of nature" (Haas 1996:27–28). Thus, according to this argument, experts not only provided technical advice, which is undoubtedly true.[5] More importantly, Haas's argument is that the legitimacy of their knowledge-claims led to political empowerment and the content of such claims, and "social laws" derived from them, shaped governance.

Testing an epistemic communities hypothesis entails three steps. First, a community of experts, privileged by its claim to authoritative knowledge in the issue area and with a shared policy enterprise that generates a set of norms that stem from a scientific consensus within the group must be identified. Given the outcome in this case, such norms would need to establish linkages between environment and development for the hypothesis to hold. Second, the group must attempt to influence the political process through the promotion of its ideas. Third, relevant actors or institutions must eventually adopt those norms over alternatives (Haas 1992b:34).

I will show that the explanation fails each criterion for success in this case, demonstrating the need to move beyond the theory and substance of this argument for a fuller explanation. First, I find little evidence that a coherent epistemic community formed around "scientific ecology" or that scientists agreed on "social laws" derived from that research program. Second, scientific communities have a mixed record in influencing policy. Science and "scientific ecology" certainly played a role in identifying environmental problems and influenced thinking in international organizations such as UNEP and the IUCN. Those organizations in turn helped to disseminate ecological concerns to state governments, NGOs, and publics. Not surprisingly, science also played a role in supplying technical knowledge that helped in the formulation of some specific policies in response to perceived crises and in the face of uncertainty. So, an epistemic community approach tells part of the story. However, it performs far less well on the core political issue of consensus on what should be the appropriate responses generally to global environmental problems. Consensus on the nature of environmental problems was often weak, particularly among hard scientists when it came to values, management norms, or specific responses in key cases. Also, most hard scientists came late, if at all, to development concerns.

Finally, and most importantly, the ideas behind liberal environmentalism simply did not originate among scientists, ecological or otherwise—a point I return to in the next chapter. In fact, it appears the causal arrow often ran the other way. Ideas around the norm-complex of liberal environmentalism appear to have increasingly influenced scientific work that feeds into global environmental research, rather than vice-versa.

Before going over the evidence for these findings, it is important to have a clear understanding of the policy process that underpins the epistemic communities hypothesis. The epistemic communities literature asserts that the success of ideas depends on whether *consensus* emerges within such groups (Haas 1992a:41). Accordingly, a high degree of consensus within the community makes the ideas it supports more likely to influence policy and to facilitate coordinated state action around such ideas. Low consensus diminishes the influence of the group and makes agreement on matters related to the community's expertise less likely. Consensual knowledge embodied in epistemic communities can help states identify interests under conditions of uncertainty, frame issues for collective debate, propose specific policies, and identify salient points for negotiation. Thus epistemic communities' activities can lead to international policy coordination and ultimately determine the content of governing norms and policies.

Haas is not alone in focusing on scientific consensus as a key ingredient of successful influence (for example, see Benedick 1991; Döös 1991). However, others, such as Döös, focus more on the difficulty in achieving such consensus given problems of observation, measurement, and prediction of human effects on the environment. Furthermore, Döös (1991:4–7) argues that scientific consensus, while important, can be thwarted because governments may encourage negative feedback loops once the political process gets underway.[6] Writings on scientific policy generally echo the point that a variety of factors in the political arena may thwart or redirect expert knowledge. Some philosophers and sociologists of science go further, arguing that the conduct of disciplinary scientific research can never be fully exempt from politics over the internal construction of knowledge and scientific activity. In this view, the broader context of scientific research always interacts with societal and, in the case of policy-related science, governmental structures.[7] Whatever one thinks of the deep critique of knowledge construction, the question should be addressed of how does the formulation of policy interact with an apparent scientific consensus? In other words, what happens in the process of scientific influence that leads not only to positive and negative feedback loops for action, but also to the kind of action that is deemed appropriate?

The epistemic community hypothesis in its pure form appears to posit a fairly linear relationship between scientific consensus and policy outcome, with only minor institutional hurdles to overcome. In other words, once a sufficient level of scientific consensus is achieved—though it must navigate through national bureaucracies, convince leaders, and respond effectively to critics—it should provide the substantive basis on which to build agreement. The literature emphasizes channels of communication and influence in national governments, identifying these as the main political hurdles. Following this logic, the influence of an epistemic community depends in part on its privileged access to officials and leaders of national administrations or international secretariats, or to its members joining such bureaucracies themselves. Membership in such organizations helps to ensure the institutionalization of ideas carried by the epistemic community and the socialization of governments to the norms promoted by the group (Adler and Haas 1992:374).

The literature as a whole is somewhat ambiguous on the necessity of community members actually becoming government personnel. However, if a community's influence stems from the legitimate authority granted to it by virtue of its expertise and its policy activities are driven by principled beliefs around the issue at hand, then epistemic communities should be considered autonomous groups from the governments they influence, with their own set of interests and priorities. According to Haas, "The members of a prevailing community become strong actors at the national and transnational level as decision makers solicit their information and delegate responsibility to them. A community's advice, though, is informed by its own broad worldview" (Haas 1992c:4). If governments manipulate the activities of such groups according to government interests, their autonomy is compromised and the analytic weight that can be attached to epistemic community influence is diminished. Furthermore, if the evidence supports a very different understanding of the policy process than that just presented, the hypothesis fails.

An epistemic communities hypothesis also goes beyond a simple argument that an influential network of interested actors promoted ideas they preferred. Otherwise, concepts from the comparative public policy and transnational relations literature would suffice. Concepts such as "policy network/community," "advocacy coalition," "issue network," or "transnational social movement organization" all identify networks of actors involved in a policy, either owing to a common interest or shared policy enterprise.[8] Haas distinguishes epistemic communities by, in addition to their shared causal and principled beliefs and common policy enterprise, their "authoritative claim to policy-relevant knowledge in a particular domain . . . based on their recognized expertise within that domain" (Haas 1992c:17).

Furthermore, epistemic communities are more than mere purveyors of consensual knowledge. The consensual knowledge literature focuses almost exclusively on the uses of knowledge during negotiations. It pays less attention to the actors that carry such knowledge, their own goals and sources of legitimacy, or how such knowledge may affect interest-definition as on ongoing process outside of negotiations (Rothstein 1984; Sjöstedt 1994). In contrast, the epistemic community hypothesis draws its power from the special status accorded to the community's expertise, which gives it legitimacy. The focus on legitimacy of ideas gives the hypothesis its causal weight and analytic strength compared to other concepts. Epistemic communities are not united simply by interests; they form around specific knowledge claims and values that, to have force, must stem from those knowledge claims.

One should answer the following set of questions in the affirmative in order to have confidence in an epistemic community explanation: Was consensual cause-effect knowledge necessary for responses or action? Did knowledge come from an identifiable network or group acting with a particular value orientation? Was such a group autonomous from state actors and were the members self-recruited? Did the group push states in a policy direction they might otherwise have not taken? And, have other intervening factors been discounted (Haas 1992b; Haas 1992a, 44–45; Haas and Haas 1995:260)?

Most commonly, scholars utilizing this approach attempt to explain outcomes on discrete issues, such as ozone depletion or whaling. Such studies can easily identify single research communities, measure community influence by following how individuals move into domestic bureaucracies, and delineate the ins and outs of all relevant negotiations. Given the timeframe and breadth of this study, however, this approach is inappropriate. Instead, as Haas does in more recent studies (which are the logical extension of the approach), I will use broader strokes to focus on key ideas and the access given to major players in the relevant scientific communities who might qualify as members of an ecological epistemic community. I will look at the influence of ideas they championed and their own activity over time to determine their influence (Haas 1996; Haas and Haas 1995). To address the criticism that I have not carefully examined the influence of an epistemic community in a particular case, I also examine some specific cases in more detail, notably those where one might expect it to perform best, such as ozone and climate change.

In addition, it might be objected that it would be unfair to argue a monocausal explanation for the broad expanse of normative evolution examined in previous chapters. I would not expect an epistemic communities explanation to perform to the same standard as in a study of a specific treaty outcome or discrete environmental problem. However, the literature does assert that an ecological epistemic community holds *the* privileged position in the broader development of global environmentalism as well. This community holds such a position because of its commitment to examining cause-effect relationships through the scientific method, its allegedly holistic approach, and its commitment to environmental preservation. So it remains a worthwhile exercise to assess whether an epistemic community was either necessary or sufficient for the normative development identified in previous chapters and, if not, then to make some observations about just what role science did play in the evolution of governing norms.

THE ROLE OF SCIENTISTS AND SCIENTIFIC ECOLOGY

Scientific Ecology

I begin with an assessment of Haas's assertion that "scientific ecology" ideas dominated thinking in epistemic communities active in international environmental research. The assumption requires critical examination to avoid the circular reasoning that if outcomes reflect some ecological concern, they do so because of consensus within the relevant epistemic community on "scientific ecology."

Contrary to Haas, an examination of the work of ecologists suggests "scientific ecology" is an unlikely candidate to form the basis of epistemic consensus and values. The problem stems first from Haas's various definitions of ecology, which conflate ecological thought and the work of scientific ecologists, who, using scientific methods, simply study how living matter interacts with its environment. As a result, his description of an ideal-type scientific ecologist mixes facts and values from different branches of ecology, other disciplines, and the environmental movement. The links between a specific set of values and what ecologists actually do, or what their findings suggest for norms of human behavior, are simply far less direct than he suggests.

For Haas (1992a:43) ecology "has been described as a framework that assimilate[s] other scientific disciplines." It does so because it studies the interaction of living (the biosphere) and nonliving realms (the atmosphere, geosphere, and hydrosphere). So ecologists, who come from a variety of backgrounds, "share a common belief in the need for an holistic analysis that is sensitive to the possible feedback and synergistic relationships among a variety of variables." This ideal-type description of scientific ecology, however, masks sharp disagreements among ecologists, not to mention among many natural scientists who study environmental problems but do not necessarily subscribe to an ecological philosophical position. Two faulty implications stem from this definition: first, that ecologists uniformly adopt a systems approach; second, that "scientific ecology" dominated international environmental discourse. I challenge each assumption in turn.

The ecology Haas describes most closely resembles ecosystem and systems ecology, the latter pioneered in the work of Eugene Odum and, to a lesser degree, his brother Howard.[9] Haas may also have in mind earlier pioneering work of influential ecologists such as Soviet geochemist Vladimir I. Vernadsky, who took a holistic approach to the subject. Vernadsky was one of the first to use the term biosphere and to stress biogeochemical cycles to understand the interrelationship between living and nonliving systems. Key elements of system ecology include its emphasis on the ecosystem concept and the flow of energy through them, the self-regulatory/functional properties of living systems, and the existence of negative feedback loops in nature. System ecology also conceives of nature as composed of "innumerable, partially overlapping systems" (Hagen 1992:131). This holistic branch of ecology did indeed influence international scientific study of the Earth's environment in the late 1960s through the International Biological Program (IBP), a large-scale transnational research program on ecosystems, although the fissure between evolutionary and systems ecology split scientific support for IBP (Hagen 1992).

That division highlights the second of two difficulties that arise with the focus on systems ecology. First, some attempts to apply its insights to social and political systems were largely discredited within the broader ecological community because they suggested an extreme version of social control that appeared anti-democratic. Howard Odum's (1971) semi-popular *Environment, Power, and Society* exemplified this trend. Early chapters on concepts of ecosystem ecology, systems modeling and the limits of industrial growth were highly regarded. However, Odum's application of his systems approach to politics and religion, and the simple control loops of his energy diagrams to explain voting, public opinion, taxes, and even revolutions and war, suggested the need for a coercive system of social control, not the democratic choice he claimed to promote (Hagen

1992:135; Bowler 1992:540). In general, systems ecology has been character-ized by a strong management orientation that makes many environmen-talists uncomfortable, especially when it comes to the global environment (Hagen 1992:138–140; Finger 1993:42; Hawkins 1993). Thus, one is hard pressed to find a consensus that the "ecological discipline . . . does not ap-pear to reflect and reproduce patterns of inequality and dominance that may exist in domestic and international society more broadly" (Haas 1992a:43). The evidence thus seems to contradict the vision supposedly put forward by the scientific ecology epistemic community as presented in Haas's writing.

The second problem, noted above, is that stark divisions within ecology arose in the 1960s and 1970s. This split does not inspire confidence that consensus existed on ecological ideas that supposedly informed thinking in transnational scientific communities. As one prominent ecologist who has worked extensively on transnational environmental issues put it, "I know a lot of ecologists and if two of them share the same perspective, I will eat my hat."[10] The deepest and most acrimonious splits occurred just when ecol-ogy saw a huge increase in numbers of practitioners, money, and interest from governments, and high expectations from the public who looked to it for insights into environmental problems.

A complete history of the split is not necessary here, but a few aspects merit highlighting. The main split came from population ecologists who challenged the holistic approach of the Odums, opting instead for a more evolutionary stance where individualistic competition determines the structure of a region's ecology and the evolution of species themselves. A series of more technical debates about the use of mathematical modeling and the like also arose. The splits were not merely disciplinary debates, but affected how ecologists saw the application of their discipline to human behavior. As one historian of science puts it: "Many ecologists accept theo-retical models that are quite explicitly opposed to the holistic perspective of the radical environmentalists. The development of scientific ecology cannot be equated with the rise of environmentalism, nor have the ten-sions between these two areas diminished in the modern world (Bowler 1992:536–537)." Or, as another author observes, "Ecosystem ecology provid-ed the ideal perspective for examining critical environmental problems, but for many evolutionary ecologists this perspective lacked an acceptable intellectual foundation" (Hagen 1992:163).

Here the problem of values must be confronted head on. True, ecology as a system of thought does imply many of the values Haas identifies. For example, Tim Hayward (1994:31–32) lists three core values of ecological thought: live in harmony with nature (humans are a part of nature, not separate from it); overcome anthropocentric prejudice; and recognize intrinsic value in beings other than humans. But Hayward harbors deep skepticism about the ability to derive these values from ecological science. "My conclusion, then," he writes, "is that the normative regulation of human affairs cannot necessarily be derived from ecological insights, at least to the extent that they follow a logic which ecology is insufficiently equipped to illuminate." Rather, an ecological lens has been attached to values that come from elsewhere (1994:34). Unsurprisingly, Bowler (1992: 536) finds that ecology has been used as frequently by the industries that some environmentalists criticize as by environmentalists themselves, each finding support within ecological science for their position—and controlled exploitation is as much a part of ecology as environmental protection. Hence, Haas's (1992a:43) ultimate claim linking scientific ecology to environmental preservation as "an absolute end" is incorrect. Given these divisions, Bowler's (1992:504) caution seems appropriate to keep in mind: "The very word 'ecological' has come to denote a concern for the environment. In science, however, 'ecology' is merely the discipline that studies the interactions between organisms and their environment. History shows that such studies can be undertaken within a variety of different value systems." My critique makes no judgment on the merit of ecological values *qua* values. It only questions the claim that they arise as the epistemic community literature suggests, based on a scientific and normative consensus among ecologists.

Finally, the following claim also has more to do with ecological thought than ecological science: "[Ecologists] do not view environmental policies in terms of opportunity costs, as some economists commonly do. Consequently, when involved in international environmental negotiations, they have encouraged behavior that is different from previous patterns of collective action" (Haas 1992a:43–44). If that were true, one would have to conclude from chapter 3 that the ecological epistemic community has been only marginally successful. While environmental protection is certainly present in the complex of norms governing international activity on the environment, the goal of economic growth for human needs, albeit a more environmentally friendly growth, remains paramount. Furthermore, environmental policies are indeed evaluated as much if not more by the costs and benefits of various actions than from the position of an environmental protection ethic as expressed above.

By contrasting environmental ecology to economics, Haas correctly points out the different values and assumptions of ecological thought and classical economic thought. The difference comes from environmental economics' basic starting point, consistent with classical economics, of valuing the environment to ensure the costs/benefits of human activity are properly considered. The basic underlying value is that "taking care of the environment is in humans' interest" (Haas 1994:102). But human interest is dependent on a prior set of values of what humans want and need, and hence comes from socioeconomic and historical circumstances. The environment is instrumental, thus important, and ought to be brought into economic models.

Ecological values, as portrayed by Haas and as commonly understood in philosophical literature, imply a much more radical position with implications for the way policies ought to be framed. That perspective gives value to the environment as an end in itself, not related to its use by humans. The contrast is important to assess critically and accurately the source of ideas of environmental governance. Otherwise, it would be tempting to equate just any concern with the environment with the influence or success of a scientific ecology epistemic community. Scientific ecology does not reflect just any concern with the environment, and stands in contrast to fundamental tenets of the norm-complex endorsed by the Brundtland Commission in 1987, and certainly of liberal environmentalism as articulated at UNCED in 1992. These findings contradict the existence of a coherent set of norms produced by policy-oriented ecological scientists with the authority of epistemic consensus.

In the absence of an ideal-type epistemic community, scientists may still influence policy. Below I identify scientific communities directly involved in negotiations or policy and conduct a process trace of their attempts to influence normative outcomes. Indeed, leading scientists at the cutting edge of global change research are generally not concerned about the philosophical or disciplinary perspective from which they come, but often simply go where the science takes them (William Clark and James Bruce, author's interviews). Admittedly, the science agenda for global environmental change as it developed in the 1980s has increasingly focused on the interaction of biological, chemical, and geophysical processes and their relationship to human activity. But for most scientists, this has little to do with a particular value orientation, ecological perspective, or conscious policy enterprise.[11] To simplify matters then, in the remainder of the chapter I focus specifically on the scientific communities most directly involved in global governance and ask first whether they represented a "scientific ecology" approach, and second what role, if any, they played in providing the foundation for international environmental norms.

The Scientific Community

Caution is warranted in determining what group of individuals might be properly called a scientific ecology epistemic community. For example, simply counting Maurice Strong, Jim MacNeill, Peter Thacher, Mustafa Tolba,[12] and others who took leadership roles in key UN environmental organizations and processes would unfairly stretch the use of the term. Even though some members of this group have science backgrounds (Tolba, for example) others do not. Moreover, some do not base their goals or values on cause-effect relationships in ecological science, even if they see the value of scientific research for environmental governance. This group, and its allies in domestic bureaucracies, might better be termed "knowledge brokers"— intermediaries between original researchers and policymakers or those involved in the policy process (Litfin 1994, 4, 37–40)—or more generically as policy entrepreneurs. Caution is also warranted when attaching analytic value to such labels, however, since "knowledge broker" still implies that the source of legitimacy for such a group rests on its use of scientific knowledge rather than the promotion of a particular set of values. While many key individuals used science to back their claims, it was not always primary in their attempts to influence others or shape the discourse around environmental governance. The epistemic community hypothesis, as I argue above, must therefore not merely focus on this group of like-minded influential individuals—indeed its value added is to provide a causal link between the authority of the knowledge claim and policy change. An epistemic community is a group with particular expertise who draw on that expertise to formulate not only technical advice or scientific research programs, but also goals and proposals that could potentially shape behavior of governments or other groups who partake in governance structures.

To locate possible members of an epistemic community, I briefly map the terrain of environmental science and international environmental policy. The range of environmental negotiations that require scientific inputs, and the number of actors and their interactions involved, make for a dense network of interactions and feedbacks that appears to make identification of an epistemic community daunting. For example, the Canadian Global Change Program (1996:91–94) identified more than 150 different research programs and organizations involved in global change activities, the majority of which involve scientific research. While global change research is currently the most prominent international environmental research program, one could make similar lists for specific concerns, ranging from big issues such as biodiversity, forestry, or ocean pollution to a myriad of specialized environmental or conservation issues on the international agenda.

Despite the plethora of actors and organizations, key umbrella institutions or groups closest to international environmental negotiations and policy processes can be identified. Among the influential nongovernmental scientific organizations are ICSU (International Council of Scientific Unions) and some of its prominent programs and member organizations such as SCOPE (Scientific Committee on Problems of the Environment) or IGBP (International Geosphere-Biosphere Programme); IUCN (World Conservation Union which includes governmental and nongovernmental representation); and IIASA (International Institute for Applied Systems Analysis)—an east-west think tank and product of the Cold War that does high-profile, interdisciplinary, and policy-relevant research on global problems. Some intergovernmental organizations have also played central roles, such as UNEP, UNESCO, and WMO (World Meteorological Organization). The latter organization became associated more closely with environmental problems as atmospheric issues gained ascendancy on the international agenda.

On any given issue, particular organizations (or scientists within those organizations) often assume leadership or coordinating roles, and organizations frequently collaborate to create specialized bodies to conduct research on specific topics. For example, ICSU and WMO collaborated to create the Global Atmospheric Research Programme and then the World Climate Research Programme. Similarly, UNEP and WMO jointly sponsor the Intergovernmental Panel on Climate Change (IPCC). Of all these organizations, UNEP and IIASA probably best represent the value-orientation of "scientific ecology," although neither has used the label. As I will show, however, these organizations, especially IIASA, were not necessarily the most successful in influencing the content of policy. In addition, both organizations have themselves adapted their research to provide a closer fit with liberal environmentalism.

With the backdrop of key organizations in mind, a process trace of the influence of scientific ideas and knowledge from these organizations (and individuals) can determine the strength of the epistemic communities hypothesis, even accepting that an ideal-type epistemic community did not exist.

THE INFLUENCE OF SCIENTISTS AND
SCIENTIFIC ECOLOGY

The evidence shows that scientists played a remarkably minor role in the lead-up to and activities of the three major norm-articulating events examined. Scientists had more success in bringing particular problems to the attention of governments and some environmental scientists helped shape ideas about international environmental governance through institutions such as IUCN and UNEP. However, even within these primarily environmental organizations, environmental governance faced pressure to respond to development concerns and these organizations were unable to develop ideas that successfully bridged that tension in order to forge a focal point for normative consensus.

Science and Ecology in the Lead-up to Stockholm

A counterfactual example helps illustrate the limited influence of science and ecological ideas at Stockholm. Had UNESCO's 1968 Biosphere Conference in Paris set the pattern for environmental norms rather than Stockholm, the epistemic communities hypothesis would find strong support. As much as any other attempt to coordinate global environmental action since, the Intergovernmental Conference of Experts on a Scientific Basis for Rational Use and Conservation of the Biosphere endorsed an ecological approach to global environmental problems and reflected concerns associated with ecosystem or systems ecology. However, it did not set the trend. Instead, many of the environmental ideas that spurred global research and raised expectations among scientists fell into the background at Stockholm, except when it came to recommendations that dealt specifically with further research itself.

The Biosphere Conference did attempt to apply ecological ideas to the development process and bring environmental concerns to developing countries (Caldwell 1990:44–45; Adams 1990:30–36; McCormick 1989:88–90). Systems ecology, with its emphasis on management of ecosystems, provided a way to move beyond traditional concerns of conservation and endangered species, which did not interest the UN Economic and Social Council (Adams 1990:32–33). The conference successfully set in motion attempts by IUCN, UNESCO, and other conservation organizations to develop ecological principles for development. For UNESCO and IUCN, though, that focus came from the new development discourse among Third World countries, not from ecological ideas themselves. Still, these efforts came to fruition in statements such as the 1973 publication of IUCN's "Ecological Principles for Economic Development" (Adams 1990:32, 143–145). Many of these ideas can be traced to a scientific ecology epistemic community involved in the IBP, some members of which directly participated in the Biosphere Conference. The IBP had a major influence on thinking there, especially its call for the establishment of an interdisciplinary and international program of research on the rational use of natural resources to deal with global environmental problems.[13] It should also be noted that the values expressed were much more anthropocentric than those put forward in the early environmental movement, and therefore represented the management orientation of systems ecology more than simply an environmental preservation outlook.

The influence of the IBP, however, did not come from a consensus on a proper ecological approach *per se*, but on the support within the IBP for a research program that fit with global environmental concerns. These concerns drew especially from Odum's work and from supporters who pushed systems ecology, at least in part because they hoped it would raise the status of ecology to match more established disciplines such as molecular biology. Hence, although it had some other elements, "for all intents and purposes, [the IBP was] an international study of ecosystems" (Hagen 1992:170). The ecological community itself split in terms of support for the IBP, as many ecologists from a non-systems perspective were suspicious of the "big science" orientation of IBP projects and its focus on ecosystem research (as opposed to evolutionary ecology, for example). Many prominent ecologists felt that large-scale ecosystem studies were not the best part of ecology (Hagen 1992:172).

The most concrete outcome of the Biosphere Conference was UNESCO's Man and Biosphere Programme (MAB) launched in 1971. Mandated to study global relationships between human activity and the environment, MAB clearly took an ecosystem management approach and many projects it sponsored linked natural ecosystems and human use in single research projects. However, older nature preservation outlooks also remained, particularly in MAB's Project 8, which created "biosphere reserves." These reserves were often renamed areas already protected, did not really reflect ecological selection criteria, and did not succeed very well in creating protected spaces in developing countries (Adams 1990:33–36).

Scientific ecology did also influence the organization of scientific research and framing of conservation concerns (especially of animals and plants) at the international level. However, during this time period, most problems in practice were still treated as discrete problems of purely nature preservation or conservation, received a relatively low profile on the international agenda, and required relatively little coordination among states. The plans that came out of the Biosphere Conference such as MAB remained largely removed from international governance (that is, the realm of state or other actors rights and obligations, regulation, and so forth). Hence, in the long run, those plans had a limited influence on governing norms.

The Stockholm Conference

Scientists played a much greater role in the Biosphere Conference than at the UN Conference on the Human Environment at Stockholm. The latter reflected some ecological ideas, but mainly demonstrated the wide disjuncture between the concerns of scientists and of diplomats (Caldwell 1990:44–45). Scientists did not provide an ideational basis for the norms articulated at Stockholm and little consensual knowledge existed among the hard scientists who served as the main advisers to the secretariat or national delegations. Finally, at both the national and transnational level, scientists involved were more often reactive than proactive in conference preparations, with only a few exceptions.

Two isolated examples illustrate the haphazard way scientists did and did not influence events. The first is the fact, pointed out in chapter 2, that Maurice Strong replaced Swiss biologist Jean Moussard as secretary-general in 1971 after it became clear that Moussard did not possess the political savvy necessary to make the conference a success. Although the reasons for the replacement go beyond Moussard's scientific background, it indicated the difficulties that scientists would face in maintaining primacy when bringing environmental concerns into the mainstream of the multilateral agenda, or shaping it.

Second is the story of Svante Odén, the Swedish scientist who almost singlehandedly convinced his government of the need for an international conference to promote cooperation on acid rain. Despite little training in atmospheric science, he successfully used his own theory to convince politicians and the Swedish people that lakes and rivers in Sweden were becoming acidic partly as the result of sulfur from smokestacks in other countries, adequate responses to which would require international cooperation. Largely as a result of Odén's efforts, Sweden proposed the Stockholm conference.

The scientific story on acid rain is telling because it worked in a way quite different than an epistemic community hypothesis would suggest. The earliest related research dated back all the way to 1661 when investigators in England noted that industrial emissions affected plant and animal health and that England and France exchanged windborne pollutants. The term "acid rain" itself dates to 1872 when a British official who monitored pollution wrote about increasing acidity of rain as one got closer to industrial centers.[14] The subject remained dormant for nearly 100 years until research programs started to examine the effects of acidity in precipitation on various living organisms. While a number of independent researchers studied related phenomena that would be brought together in later research programs, the interest of policymakers was a result mainly of the efforts of Odén, a soil scientist and adviser to the Swedish government, who also happened to star in a popular television show. Odén was the first to publish a complete theory of acid rain, in which he linked air pollutants containing sulfur and nitrogen to increased acidity in rain that fell large distances away from their industrial sources. He also identified ecological consequences of acid rain that ranged from changes in the chemistry of lakes to accelerated damages to materials. Interestingly, Odén first published his theory of acid rain in 1967 in a prominent Stockholm newspaper, not a scientific journal (although the next year he published an article in *Ecology Committee Bulletin* that stimulated interest in the scientific community).

As a result of his work, Sweden sponsored a scientific study to try to determine the extent of the problem. It presented the study at the Stockholm conference, which then acted as a catalyst for expanded research programs in other countries. Odén's success stemmed from his personal access to policymakers, entrepreneurial efforts to spread word of his theory, and public popularity, rather than scientific consensus. This story, although it is but one example, lends little support to an epistemic communities hypothesis, which emphasizes the need for consensual scientific knowledge.[15] Instead, it illustrates the nonlinear relationship between scientific knowledge and political action on the problems such knowledge addresses.

In terms of the broader issues of governance, the normative compromises that arose at Stockholm came largely independently of scientific input. Ideas that framed the conference stemmed mostly from the entrepreneurial leadership of conference Secretary-General Maurice Strong who brought together strands of environmental and development discourse. Ultimately these compromises were political, but were facilitated by Strong's organization of two key meetings prior to the conference. These meetings helped forge a consensus among developing country economists that environmental protection could be accommodated, at least to some degree, while still making economic development a priority.

The first meeting could be considered as loosely comprising members of a scientific ecology community. The second, discussed in the next chapter, comprised members of the development community who had some interest in the environment (with some overlap). Strong himself ranks the latter meeting as being the more significant of the two in shaping the Stockholm agenda (author's interview). Even the first meeting did not involve ecologists as such, but people with related technical expertise who were sympathetic to the broad goals of ecology. The main link with ecology among the group was systems theory, which some ecologists had adopted. Three members of this meeting had been involved with the MIT team that worked on *Limits to Growth*, which used computer simulations grounded in systems theory.

Strong called the meeting shortly after his official appointment took effect in January 1971. The meeting of a small group of experts at MIT mainly concerned environmental issues (see chapter 2 for a list of participants), although Strong also wanted to link issues of environmental control with economic development in both industrial and developing countries. Strong (author's interview) said a major theme was to move thinking about the environment beyond a simple concern with pollution to a view that looked at industrial society and its effects on the environment in more systemic terms. As indicated in chapter 2, however, the meeting mainly emphasized environmental protection and a conservation ethic that emphasized sustainability for the sake of future generations.

One scientific meeting did directly address the relationship between environment and development—a SCOPE working party meeting in Canberra, Australia (Aug. 24–Sept. 4, 1971).[16] ICSU (1969:25–32) set up SCOPE in 1969 to report on the "problem of the human environment that humankind is altering." The report by a working party on basic environmental issues in developing countries, composed mostly of scientists from Africa, Asia, and Latin America, stressed the importance of what it called an "ecological approach" to environmental problems (SCOPE 1972). This approach would emphasize determining the "carrying capacity" of ecosystems, which depends both on plant and animal species, and the socioeconomic values of the society. It thus supports the incorporation of ecological concerns into development planning decisions and owed much to ideas present in the IBP and that came out of the Biosphere Conference.

Strong attended the meeting, but said it had less influence on the conference secretariat and the agenda of Stockholm than the two meetings already mentioned. As Strong saw it, "the Canberra meeting was more of a scientific meeting discussing the broader issues in scientific terms, not so much trying to write the Stockholm agenda" (author's interview). The Canberra meeting mainly aimed to bring Strong and the secretariat up to speed on the scientific issues. Strong had been looking for a source of scientific advice and SCOPE fit the bill. So, he paid for the meeting after discussions with SCOPE Secretary-General Tom Malone (an American scientist), and SCOPE became a key source of scientific advice for the conference (R.E. Munn, author's interview).

However, SCOPE did not directly address problems of environmental management—virtually no social scientists were involved, and, although SCOPE aimed in part to bring together individuals from different scientific unions, consensual knowledge was not a key component of the exercise. Scientists from different disciplines often talked past one another, although some learning did occur as ICSU and SCOPE scientists tried to understand the language of scientists involved in IBP, for example (R.E. Munn, author's interview). Indeed, some scientists came to SCOPE from the IBP or MAB, although turf wars between government and nongovernmental scientific programs were not uncommon.

Learning about development was more difficult, however, for many individuals within SCOPE. Most, including those in leadership roles, were (and are) hard scientists, uncomfortable with policy questions at all, and with a weak understanding of development concerns. Attempts at dialogue often went poorly because, as one participant put it, "the Third World and the hard scientists from Europe and North America hadn't talked to each other and they didn't really understand each other's problems at all" (R.E. Munn, author's interview). That lack of understanding might have contributed to the limited influence of SCOPE in conference outcomes. Hence, although the Stockholm Plan of Action echoed many SCOPE proposals on research and education, the ecological approach was buried under the concern of developing countries for economic growth.

In official preparations and at the conference itself, science played a role mostly in technical matters, and this did help promote interest in environmental concerns. For example, a number of analysts have noted that the preparation of country reports on the environment had a positive influence on government interest in the environment and helped build domestic infrastructure, particularly in developing countries (Engfeldt 1973). However, participating scientists did not constitute an epistemic community as such. Neither did the conditions identified by the hypothesis, such as scientific consensus, appear to be important for the impact they did have. In many cases, national scientists were brought into the process by governments, not vice versa. As one Canadian scientist put it:

The scientists in Canada I know took it as a big pain in the neck when [government] organized all these task forces and committees and it just meant that people couldn't get on with what they wanted to do. So they were drafted into these things. In fact they used to send petitions around that all political meetings like that should be canceled for 10 years (R.E. Munn, author's interview).

Scientists also influenced discussions on the nature of environmental hazards and on a number of specific recommendations in the action plan.[17] Scientists were most successful with specific proposals on their own turf, such as with a SCOPE report that led to the creation of the Global Environmental Monitoring System (GEMS), a component of the Earthwatch system that was one of the most significant concrete results of Stockholm (SCOPE 1971). These activities likely facilitated the strengthening of transnational environmental science research networks and the names of prominent scientists who participated in activities related to Stockholm often appear in later ICSU or other transnational research programs or activities.

However, as the conference preparations got under way, diplomats gradually took over from scientists in the formulation of policies and framing of global environmental problems. According to Lynton Caldwell (1990: 62), a long-time observer of and participant in UN environmental diplomacy, "There was conflict between science advisers and foreign policy advisers at Stockholm reflecting differing assumptions regarding the bases and priorities of international cooperation. These differences . . . were never wholly overcome."

The most active scientists thought Stockholm should promote and institutionalize a planetary conservation ethic that would transcend national allegiances. Such a position supported the creation of mandatory rules in international law that could be enforceable directly on individuals and transnational corporations (Caldwell 1990:42). However, the realities of international law and politics militated against this approach because it conflicted with sovereignty and rules that largely left the regulation of individuals and corporations to national laws, and governments were unwilling to relinquish freedom of action and control over domestic development. Scientists were particularly disappointed by the unwillingness of delegates to take a holistic approach. Instead, they displayed an overriding concern with maintaining sovereignty and what scientists perceived as short-term interests.

Perhaps that in part explains why IUCN downplayed the importance of the conference, even though it actively participated in the preparatory committees. The significance of the new perception of the position of developing countries in relation to environmental concerns was lost on IUCN, which in its annual review listed with enthusiasm the achievements of the Second World Conference on National Parks and adoption of the World Heritage Convention before its lukewarm appraisal of Stockholm.[18]

As a group, scientists felt disappointed with what they had achieved at Stockholm. The then executive director of the U.S. environmental committee of the National Academies of Sciences and Engineering summed up scientists' impact this way:

> Despite this promising start [in preparatory activity] science never emerged in Stockholm in the role of a recognized, let alone equal, partner in a common enterprise. Although there had probably never been an intergovernmental conference in which science was accorded a larger and more direct share in the preparation as well as the outcome, this was hardly apparent during the Stockholm proceedings (Kellermann 1973:485).

Scientific Ecology and International Environmental Governance: 1972–1987

The creation of UNEP and Maurice Strong's leadership of it meant the ideas and scientific proposals that came out of Stockholm would immediately find a comfortable home within the United Nations. For example, Strong immediately provided money for GEMS proposed by SCOPE, a program UNEP continues to fund.[19] Strong's leadership also assured that the overall norm-complex of Stockholm, which included the uneasy mix of development concerns and state sovereignty in addition to environmental protection, would also guide scientific research and the framing of environmental issues.

Perhaps the high point in the influence of scientific ecology on this agenda came with the attempt to entrench the idea of "ecodevelopment" within UNEP. This attempted "marriage" of ecology and economy did not come from the science of ecology, so it could not really be said to have arisen from an epistemic community in its purest form, but it did attempt to stay close to the ecological values of systems ecology. While ecological ideas clearly influenced ecodevelopment thinking, ecodevelopment did not translate into great success in shaping governing norms because it sold poorly in developing countries and could not forge a broad-based normative consensus.

Note also that Maurice Strong, who coined the term while head of UNEP, said the single biggest influence on his thinking about ecodevelopment was Ignacy Sachs, a French development economist, whom Strong credits with giving the concept intellectual content.[20] I do not mean to suggest that UNEP did not try to incorporate ecological principles in its programs. For example, its regional seas program certainly did (Haas 1990), but the focus on ecological principles tells only part of the story. The main orientation of efforts to shape how the international community would respond to environmental threats stemmed from the attempt to mix ecology and development, which did not come mainly from a scientific ecology community.

Furthermore, UNEP as an organization had only limited success selling ecodevelopment in the developing world, because it was perceived as laying too much emphasis on the ecological side of the agenda.[21] UNEP could not forge the necessary North-South consensus or alliances among key state or institutional actors to create a normative focal point for environmental governance.

The World Conservation Strategy attempted to use the term "sustainable development" to overcome these difficulties, but had only a small impact on overall governance structures, further demonstrating the limitations under which policy-oriented scientists and environmentalists worked. The final strategy focused on conservation of living resources, although some sections did mirror ecodevelopment thinking.[22] As chapter 2 indicated, at bottom WCS argued that development, since it alters the biosphere, must take conservation into account to be sustainable. The solution, then, is to give conservation a higher priority.

The WCS undoubtedly contained some ecological ideas, though it did not fully overcome its conservation orientation. That the final strategy focused as much on development as it did resulted from consultations and negotiation on a second draft among delegates from developing countries at the IUCN 1978 General Assembly in Ashkhabad, USSR. The revised draft then went through consultations with UNEP, WWF (its co-sponsors), FAO, and UNESCO (McCormick 1989:162–170). Despite the compromises and influences from these various constituencies, it never overcame its lack of attention to the main concerns of developing country governments, nor did it take into account the essentially political nature of development nor the social production of nature (Adams 1990:50–51). The problem stemmed in part from the difficulty IUCN leadership had with its own constituency in getting development on the table. As a result, rather than recasting the debate, it tacked development concerns onto the traditional conservation agenda. The greater the degree to which ecological ideas dominated that seemed insensitive to the above concerns, the more WCS seemed to reflect old environment and conservation thinking of the 1970s to many analysts (Adams 1990:50–51; McCormick 1989:165). The perceived ecological focus decreased the likelihood that WCS would make a long-term impact. Ecological ideas had to be substantially recast before they could provide a pillar for international environmental governance.

Science, Ozone, and Global Ecology

While environmental- and conservation-oriented international organizations struggled to find the proper meshing of environment and development concerns, some transnational scientific communities did achieve major successes in raising the profile of global environmentalism. In particular, the rise in prominence of atmospheric science research programs—although many were U.S.-based and not transnational until much later—helped raise the alarm over the threat of ozone depletion and climate change.[23] UNEP, WMO, and ad hoc intergovernmental bodies set up to study and report on ozone depletion also played important roles in promoting international action and raising the profile of global atmospheric science.

A number of analysts point to the ozone debate and eventual action as a defining moment in the shift to truly global environmental concerns and have commented on the advent of global ecology starting around this period (Sachs 1992, 1993; Hawkins 1993). At least one author (Finger 1993) attributes this trend in part to the influence of atmospheric chemists, geologists, oceanographers, and climatologists who helped define a new type of ecology called "global ecology." Whereas atmospheric scientists, climatologists, and other scientists involved in global environmental issues gained higher profiles and larger research budgets as attention turned to global environmental concerns, the question still remains whether they actually shaped how such problems would be addressed. The evidence shows that even in most-likely cases—such as international action to prevent ozone depletion—an epistemic community hypothesis achieves only mixed success.

Most observers consider the Vienna Convention and subsequent Montreal Protocol to combat the depletion of stratospheric ozone the result of the most successful diplomacy to date to solve a highly technical international environmental problem. The case is a most-likely candidate for an epistemic community explanation since knowledge about the problem, its sources, and detection depended wholly on advanced science. Yet scientific consensus did not appear to be a major factor in getting political action; scientists themselves did not seem to push for a clear set of ecological values; and scientists virtually ignored the North-South dimension of the problem which became crucial to the long-term success of the treaty and to the broader normative structure of environmental governance to which the ozone issue contributed. For example, scientists were not important actors in introducing provisions to allow longer phase-in times for developing countries or in the process that led to the 1990 London Amendments to the Montreal Protocol, which created the Multilateral Ozone Fund to help developing countries phase out ozone-depleting chemicals.

Only rarely does a community of experts first establish consensus and then mobilize public policy as the epistemic communities hypothesis suggest. More typically, research programs on an issue will go on independently until a handful or even one entrepreneurial piece of research (or individual researcher) manages to start putting theories together, garners media attention, or catches the ears of policymakers. The common pattern is then that the initial flood of public attention produces more money for research and only then may a scientific consensus begin to build (Kowalok 1993).

In this regard, the story of getting the attention of the U.S. government on ozone is not dissimilar to the acid rain story mentioned earlier, with the exception of its relative quickness—scientific research about human-induced ozone depletion only started around 1970.[24] At that time, large-scale research concentrated on the potential threat of supersonic transports (SSTs). Some scientists feared that the release of nitrogen oxides and water vapor might deplete ozone in the stratosphere, the protective layer of atmosphere 10–50 kilometers above the Earth's surface where most ozone is found. This ozone layer protects the Earth's surface from ultraviolet radiation. These concerns led the U.S. Senate to terminate funding for a planned fleet of 500 SSTs (a Boeing project), despite a prominent study that concluded that ozone depletion from SSTs would be insignificant.[25]

Large-scale transnational research also occurred, but concern quickly shifted to the ozone depleting potential of chlorine as SST programs were cut or scaled back. The new concern came from research by NASA scientists on possible effects of the space shuttle's rocket boosters, which would directly inject chlorine into the stratosphere. However, for political reasons, NASA scientists downplayed the threat from the shuttle and instead emphasized the threat from volcanoes, which left some scientists puzzled as to the concern since they did not see the danger for lack of a major source. Then, in 1974, F. Sherwood Rowland and Mario J. Molina's famous article in *Nature* identified chlorofluorocarbons (CFCs) as plentiful sources of atmospheric chlorine. Thus, a source of concern had been found, and an entirely synthetic one at that.

What got the ear of government, however, was not activity by scientists *per se*, but a front-page article in *The New York Times* following a meeting of the American Chemical Society where Molina and Rowland presented their data. Their paper warned that predicted ozone loss could lead to significant rises in the incidence of skin cancer and possible crop loss. The *Times* article was followed by extensive national television coverage that prompted more letters to Congress than any other issue since the start of the Vietnam War (Brodeur 1986:70). After the public outcry, the U.S. government funded further large-scale and coordinated research efforts. What happened next in terms of policy responses is well documented elsewhere. For the purpose here, the significant pattern in U.S. regulation (starting with bans on the use of CFCs in nonessential aerosols) is that it consistently outpaced scientific consensus on the extent of the problem until the late 1980s. Only then did the transnational scientific consensus on the causes of ozone depletion became more solidified.

However, it should also be noted that some government scientists did promote a precautionary stance despite uncertainty, at odds with the chemical industry. For example, Russell W. Peterson, chair of the President's Council on Environmental Quality in 1976, and formerly a chemist with DuPont for 26 years, argued that despite equivocal science, "chemicals are not innocent until proven guilty." He then argued for restrictions on CFCs (Brodeur 1986:74). However, the U.S. government acceptance of a precautionary stance waxed and waned depending on the leadership within key government bodies such as the EPA (Brodeur 1986; Litfin 1994:61–73).

At the international level, the ozone issue represented a defining moment in global environmental cooperation and seemed to demonstrate the ability of science to influence global governance more generally. The interaction of science and policy in reaching agreement on the Montreal Protocol has also been well documented elsewhere (Litfin 1994; Rowlands 1995a; Haas 1992b; Parson 1993; Benedick 1991). Here, I will only comment on those findings as they relate to the question of whether epistemic consensus was the main driver of policy in terms of both getting action and the type of action received. I will also discuss the broader implications of the community's activities for governing norms more generally.

No author questions that science played an important role in creating an international ozone agreement since expertise was a necessary condition for identifying a decrease in concentrations of invisible stratospheric ozone. In this sense, the epistemic community literature is certainly correct that knowledge of environmental problems is based on a scientific rationality view of the world. The ozone layer "is only available as an object of knowledge because of our scientific culture" (Yearley 1992). However, the influence of scientists on policy did not work in the way an epistemic community explanation would suggest. Neither did an epistemic community influence general governing norms, with one important exception: the Precautionary Principle. Even in the latter instance, the principle did not stem from cause-effect relationships inherent in the study of ozone depletion.[26]

Rowlands, for example, found some correlation between the level of consensual scientific knowledge on ozone depletion and international cooperation, but also noted a major anomaly in that the major international study credited with producing transnational consensus—the Ozone Trends Panel Report—did not appear until *after* agreement on the Montreal Protocol. The epistemic communities hypothesis suggests the former is a precondition for the latter (Rowlands 1995a:89). Parson (1993:60) is more blunt: "it was not science, but bargaining, that determined the decisions adopted in Montreal. The 50% cut that was agreed to had no particular scientific prominence. Indeed, the distribution of expert opinion at the time seemed strongly divided." Litfin (1994), in the most detailed test of an epistemic communities hypothesis on ozone to date, similarly argues that only after the treaty was signed did scientific consensus emerge, especially on the causes of the ozone "hole" over the Antarctic, which became known in May 1985. Litfin is convinced that the "hole" played a major role in framing subsequent negotiations, even though delegates agreed not to consider the evidence or its cause (1994:96–102). Prior to that point, scientific data had been open to wide interpretations in terms of policy implications.

However, scientists in the U.S. EPA at a June UNEP/EPA conference in Leesburg, Virginia 1986 successfully moved the "terms of the dominant discourse toward precautionary action." Although they did not promote any particular policy option, they moved the discourse by emphasizing the long atmospheric lifetimes of CFCs and long-term modeled predictions (Litfin 1994:91–92). Ironically, the discovery of the ozone hole, which atmospheric models had not predicted, rather than undermining scientists' legitimacy, strengthened the precautionary discourse because the models could no longer be relied upon. That uncertainty changed the framing of the scientific issue itself from one of ozone depletion to one of increasing concentrations of chlorine (which could be calculated without the use of atmospheric models, based on production data and atmospheric lifetime).

Once framed in terms of chlorine-loading, a phase-out no longer seemed like a drastic proposal.

Litfin thus concludes that the scientists were not the driving force. Rather, the framing of the issue by "knowledge brokers" such as EPA administrator Lee Thomas (a career bureaucrat trained in psychology) played more of a role. Thomas felt the risk and uncertainty of the ozone problem warranted a precautionary approach, a view driven by his orientation to risk, not science. Thomas successfully pushed this view over that of other officials with different orientations to risk, such as White House science adviser William Graham:

> Graham looked at it from a purely scientific perspective, whereas I looked at it from more of a policy perspective. Where there was uncertainty, he thought we needed more research, and I thought we needed to be cautious. We just looked at the same thing and came to two different conclusions.[27]

Hence, even in the case of the precautionary approach, which did find its way into broader international governance, the values could not be directly derived from the science itself. In addition, NGOs probably played an equally prominent role in promoting the precautionary approach to ozone (and more broadly). For instance, groups such as the U.S. Natural Resources Defense Council threw their support behind the chlorine-loading approach because it would lead to complete phase-outs.[28]

Nonetheless, prominent scientists did promote the precautionary principle here and elsewhere and should be credited in large part with giving legitimacy to this norm at UNCED. A prominent example of a highly respected scientist who easily crosses over to policymaking is Robert Watson, a NASA scientist who chaired the 1986 WMO/NASA ozone assessment and replaced Bert Bolin to head the Intergovernmental Panel on Climate Change in 1997. Speaking about the implications for policy of ozone depletion, Watson emphasized that his main policy concern was the long time frame to reverse effects, "which means you could not wait for cause and effect to be fully established."[29] He added that the same concern applies to the issues of biodiversity and climate change. However, these considerations clearly go beyond science, as they concern uncertainty itself and the implications of that uncertainty for economic conditions, security, and health. As Watson put it, "In all of these cases [ozone, biodiversity, climate change] it is an issue of how you make a policy judgement with significant scientific uncertainty. You simply can't wait for all the information to come in."[30] Even though this orientation to risk did not stem from scientific findings, it is fair to say that precautionary concerns did gain prominence as the result of the community of scientists involved in atmospheric issues; hence in this one respect there was an epistemic community that seemed to play a necessary leadership role.

Ironically, the success of an epistemic community on this norm came in a case where a lack of consensus on standards of validity is precisely the issue at stake: the view of science that underlies the precautionary principle remains highly controversial within the community of environmental and ecological scientists. While it is fair to say that the principle gained the status of a norm, as defined earlier, within environmental governance, the conflict within communities of environmental scientists makes it at least questionable whether the acceptance of the norm stems from consensus within an epistemic community rather than a combination of promotion by a wide variety of actors and organizations concerned about environmental impacts.

Precaution rests on a scientific basis only if one questions the adequacy of traditional statistical standards of significance to capture cumulative or anomalous, though potentially harmful, effects in the field (see endnote 26). Hull (1999) suggests, though, that the group of scientists who take this position is still in the minority. In a study that documents these two views of environmental science, she notes that experimental methods deriving from a formal philosophy of science still dominates the discipline, and practitioners from this perspective often consider non-laboratory methods "unscientific." These environmental scientists would be less sympathetic to the precautionary principle as a guide to policy because they believe that chemical causes of toxicity or harm can only be determined through laboratory tests of cause-effect relationships that adhere to the strict scientific standards required in formal positivist philosophies of science. Conversely, scientists who engage in epidemiological or "ecoepidemiological" research are more skeptical of laboratory results, noting that requirements such as linearity, replicability, statistical significance (with a focus on avoiding type 1 error, or false positives), and specificity are inappropriate for complex interactions of organisms and their environments in the field, where it matters most. In the field, relationships between chemicals (for example) and their consequences on organisms are just as likely to be nonlinear (with threshold effects, for example), nonreplicable (to the degree that ecosystems are not easily replicable under controlled circumstances), to lack specificity (individual causes may not be directly linked to precise effects), or to be missed by traditional statistical tests which may not capture rare or subtle though real effects, especially on individual organisms, owing to the variability and complex interactions of biological organisms and their environment.

The latter view clearly fits more with a scientific ecology ethos as defined by Haas, but current policy debates suggest that "science" as a legitimizer of norms is being used to support both positions. For example, the current debate about genetically modified organisms that played out in the negotiation of the Biosafety Protocol, and is likely to remain contentious within the WTO and other trade agreements, focused heavily on whether the Precautionary Principle or "scientific" proof should be the standard for limiting trade. Currently the WTO, under GATT article XX on exceptions for health and safety concerns, demands a standard of "sufficient scientific evidence" which appears to rest on a formal, deductive, and physicalist view of science where standards of proof of cause-effect relationships can rest only on laboratory experiments, although the WTO Appellate Body decision on the Beef Hormone dispute between the EU and Canada and the United States, discussed further in the concluding chapter, uneasily bridges the two positions.

As for ecological values more broadly, no one I interviewed in leadership positions in global environmental research or policy indicated that a set of ecological values *per se* was widely accepted in transnational scientific communities, and those that did hold such values may have also valued other goals, such as economic growth. For example, in Watson's view, the ozone issue is important because it potentially affects the quality of life, but his policy orientation does not stem from science itself. In an interview in 1996, he responded this way to a question about what values motivated him on the ozone issue:

> What we need to strive for is a high quality of life and within that it means good economic performance and a clean environment—I believe, just as our Vice President [Al Gore] believes, you can have good economic growth and environmental protection. You have to handle them very carefully and together and one cannot be the afterthought of the other.[31]

Watson expressed this view well after the signing of the Montreal Protocol and the Brundtland and Rio processes. If anything, it suggests a learning process might have occurred among scientists as a result of those events.

Science and the Brundtland Commission

Whereas atmospheric science contributed somewhat to raising concerns about global environmental issues, the Brundtland Commission told the world how to think about them. The Brundtland Commission process, however, only reinforced the limited influence of the ecological scientific community on governance. Had the project of preparing a report on international environmental action to the year 2000 not been taken out of the hands of UNEP, science might have played a more central role, as it appeared to in UNEP's parallel report (discussed in chapter 2).

As for the role of scientists in the WCED process itself, analyses and interviews suggest that scientists were neither the initiators nor the driving force behind most of the recommendations, perhaps due to a mandate that focused more on values than physical realities. In particular, the conclusion that economic growth is needed and will not damage the environment did not come from scientists nor was it based on a consensus on cause-effect relationships (Timberlake 1989).

But perhaps the most telling anomaly for an argument based on the influence of a scientific ecology epistemic community is the lack of influence of IIASA on global norms. This lack of influence provides another powerful counterfactual example of the weakness of the hypothesis in this case (that is, its positive influence on outcomes would have provided strong evidence for the hypothesis). The Austrian-based institute sponsored the epitome of Haas's version of a scientific ecology research program and contained nearly the ideal of a high-level transnational community of scientific experts. In 1982, IIASA undertook a well-developed research program called "Ecologically Sustainable Development of the Biosphere" which brought together historians, engineers, geographers, environmental scientists, economists, management experts, and policy people, to examine how to manage the interaction between development and environment.[32] The group involved was truly transnational owing to IIASA's stature as one of the only places where natural and social scientists from east and west interacted; it conducted policy relevant and cross-disciplinary work; and William Clark, who headed the program, saw IIASA as perfectly suited to take advantage of burgeoning research from many sources and countries around global environmental problems. Clark describes the potential he saw for the IIASA this way:

The IGBP [now at the center of global change research] . . . was beginning to take form at that time and there were other ventures internationally. It just seemed to me that there was a niche . . . for a couple of years in which IIASA might really be able to be a forum in which the relevant natural and social sciences and the relevant countries, at least of the developed world, could jointly participate in trying to sort out what the research agenda underlying this notion of what we today see as sustainable development might be (author's interview).

Clark, incidentally, is about as close an approximation to Haas's ideal-type "scientific ecologist" as one could get. He had training in ecology, his research experience ranged from detailed studies of rural development to regional ecosystems, and he had participated in a large-scale study with natural scientists on carbon dioxide, energy, and climate change. IIASA attracted him because of its systems approach. That meshed with his belief that the issue of climate change, and global change research more broadly, was "so coupled to other issues of human development and other environmental issues that the emerging notion of what is now called sustainable development" could be developed there. He felt IIASA's strengths made it an ideal setting "to do global environmental issues in an integrated way" (author's interview).

However, Clark objected to the word "ecology" as a description of the sustainable development of the biosphere program at IIASA, insisting that the word "ecologically" in the title of the program was a "bureaucratically imposed modifier" by IIASA management who represented sponsoring countries, not the work of the scholars involved. IIASA management feared the program branched too far into social issues, which they felt was "inappropriate." Clark did not share this concern since his entire project was meant to explore the "lovely ambiguity of the phrase sustainable development," which meant that the interaction of society and the environment could be explored from both the natural and social sciences. He did not see it as a project centered in the discipline of ecology, except in the broadest sense of looking at the problem of sustainable development in the context of an integrated social-environmental system. The specific influence of ecology came only from notions he borrowed from his thesis adviser, ecologist C.S. Holling, such as "surprise," "bounded stability," or "threshold effects." Such concepts are also linked to chaos theory, such as in its discussion of how ideas around small events can lead to large, unexpected changes.[33]

Given this focus, the IIASA program might seem a logical place to look for a broad scientific basis on which to frame the Brundtland Commission report. Similarly, one would expect that the team at IIASA might have sought out the Brundtland Commission if it acted in the way expected of an epistemic community representing a global change research program. However, neither occurred. Asked if IIASA had much interaction with the Brundtland Commission, Clark responded:

> I say with embarrassment, no, there was very little: We vaguely knew that the Brundtland Commission was working away. . . . In ways that it's hard for me to understand now how we could have been so unconnected. Not only was I not particularly aware of the details of what was going on, most of the scholars I engaged in the project weren't . . . at that time we were obviously invisible to them—hardly surprising—and they were pretty invisible to us (author's interview).

Despite the direct relevance of this research, the Brundtland Commission only spent one day consulting with the project and at least some people at IIASA had difficulty relating to the concept of sustainable development as framed by the Commission. In the words of Ted Munn, one of the project's lead researchers and one of the few natural scientists there who had been involved with UN processes previously and after: "We thought that as a North-South exercise it [WCED] didn't have much to do with us, I guess. And the Brundtland report was not a prescription for action, it was rather a mindset or a dream of what might be without telling anybody how to get there. So it didn't bother me or anybody I know at all. It sort of operated on a different plane" (author's interview).

Clark said that at the time there was very little communication between the community of researchers involved in burgeoning global change research programs and the more political activities that emerged from the UN system. Those who worked on the UN-sponsored Brundtland Commission were "a very different line of people" than the scientists who tried to move environmental research toward an integrated global change research program: "We were just different people. And I think it has been in large part due to the Brundtland group that those linkages are stronger today" (author's interview). Thus, Brundtland was the catalyst for bringing the work of scientists from organizations such as IIASA toward concerns of global environmental governance, not vice-versa as the epistemic community hypothesis argues.

The beginning of the major turn in environmental norms toward a more growth-oriented, Keynesian-style global management norm-complex seemed to occur largely independently of scientific ecology research programs. That is not to say that scientists did not provide technical information, nor that science or ecological ideas did not influence various recommendations of the Brundtland Commission. However, a unified scientific community did not appear to play a primary role in this turn of environmental governance and the weight of the Brundtland report did not draw its content, legitimacy, or underlying value-orientation from the findings of a scientific ecology epistemic community.

Finally, another development following the Brundtland Commission report suggests that it will be increasingly difficult for epistemic communities to fulfill the requirement of relative independence from government interests. After 1987 and the rebirth of environmentalism on the international agenda, this time as a truly global concern, governments appeared to make a more concerted effort to rein science in rather than allow "free wheeling" scientists to dictate the environmental agenda (James Bruce, author's interview). In looking at the lead-up to UNCED, it appears that as global environmental research picked up steam, the fitting of such research into palatable forms of global governance became a prime concern of major states from both the developed and developing worlds. One finds systematic attempts—most notably in climate change, but in other issue areas as well—to retake control of transnational research endeavors and the processes by which these projects feed into international policy formation.

Science and UNCED

Transnational environmental research by 1992 was better funded and better organized than in the years before Stockholm, yet a remarkably similar pattern of limited scientific influence characterized UNCED. Scientists did of course play some role. Transnational and international scientific organizations fed into many aspects of UNCED preparations and more than 160 countries submitted reports on the state of their environments, although experts other than scientists contributed to such reports on environment and development in each country or territory.[34] Individual scientists also played a role in UNCED, some serving as members of delegations and as participants in preparations of conference documents and agreements. Nonetheless, like Stockholm, formal scientific community involvement in UNCED was relatively small as professional diplomats and administrators dominated the UN negotiation process (Marton-Lefèvre 1994; Haas, Levy, and Parson 1992:33 fn. 11). Scientific knowledge was requested and supplied, but the process shaped how science would be used, not vice-versa. As the executive director of ICSU put it, Agenda 21, the blueprint for environmental action into the next century, "can be viewed as an instrument through which scientific knowledge was transformed into a uniquely UN frame of reference" (Marton-Lefèvre 1994:171).

The most direct input of the scientific community came from ICSU, who Maurice Strong invited to serve as the conference's official scientific adviser. Its main recommendations came from a November 1991 conference on An Agenda of Science for Environment and Development in the 21st Century in Vienna (ASCEND 21).[35] Although participants read like a who's who of transnational environmental science, the report came too late to have much influence in UNCED preparations, which were already well underway. In fact, ASCEND 21 took place two months *after* the original deadline for chapter 35 of Agenda 21 on "Science and Sustainable Development." The deadline was pushed back so the authors could revise the chapter based on ASCEND's recommendations. Other outcomes related to science included Agenda 21's chapter 31 on the "Science and Technology Community" and the Rio Declaration's Principle 9, which promotes cooperation to increase national scientific capacities and exchanges of scientific and technical knowledge "for sustainable development."

The ASCEND 21 report did express some policy positions, but mainly focused on recommendations related to implementation of research and observation programs; strengthening of interdisciplinary research and communication among the natural, engineering, health, and social sciences; building links between science and development agencies; and building scientific capacity in North and South (Dooge et al. 1992:5–11). Development concerns were not well integrated into the overall report and were largely ignored in the invited papers. One participant suggested that the problem in part stemmed from the difficulty most hard scientists at the conference had understanding how to relate development concerns to their work. Twenty years after Stockholm, ICSU still had little interaction with social scientists.[36] Thus, ASCEND 21 hardly represented a consensus on science and development except in the loosest sense of the word. The document suggests few participants thought deeply about this relationship.

Although ASCEND 21 recommendations stuck closely to the promotion of scientific activity, some policy-relevant themes stood out such as a focus on population and carrying capacity, consumption patterns, and a strong endorsement of the Precautionary Principle. The latter was seen as the proper response to the complexity of the Earth's systems and the uncertainty of the effects of human disturbances (Dooge et al. 1992:6–8). The issue of Northern consumption did receive attention during negotiations and developing countries successfully negotiated for the issue to be included in several chapters of Agenda 21. However, the final wording remained vague, with developed countries agreeing only to "take the lead in achieving sustainable consumption patterns" but not agreeing to specific proposals, targets, or mechanisms (energy efficiency guidelines, for example) to achieve the goal. Population had not been included in the original mandate of Rio and only made it onto the agenda at the behest of industrialized countries who wanted it paired with consumption issues.

In the end, neither North nor South seriously negotiated on bargains over population and consumption patterns, and the United States especially resisted any discussions on consumption. The debate that did occur (mostly on consumption patterns) was politically charged and produced little concrete action that drew from ASCEND's work. ASCEND could hardly be credited with having influenced this debate since the G-77 had long used the strategy of shifting international environmental negotiations from a focus on population growth, which the North emphasized, to a focus on consumption patterns in the North. The South had also explicitly made this strategy a part of its negotiating position for Rio from the start.[37] ASCENDS's support of the Precautionary Principle had much greater impact as the principle made major inroads, although it had already become prominent in the ozone and climate change negotiations.

More generally, UNCED experienced the same uneasy relationship between science and policy that pervaded Stockholm. The scientific community appeared either too unprepared, unwilling, or unable to communicate effectively within the diplomatic setting of the conference. As a result, its message often got watered down or else became one of a myriad of nongovernmental voices with no particular special status. Susskind (1994:66–81), for example, argues that these difficulties, among others, are typical of the impact of scientists on international environmental negotiations in most cases.

As in the case of Brundtland, the specific example of IIASA also deserves special mention since Strong had hoped it would play a major role (1992:22). But, according to Strong, that influence never materialized:

IIASA had an opportunity to play a special part and they did not do it. They were a disappointment, to be perfectly honest. I've always been very convinced of the systemic nature of these issues and was trying to design in UNCED a framework in which the systemic nature of those issues could be clearly seen by policymakers, and also the points of effective intervention identified. . . . I thought that we had a great opportunity to demonstrate this. . . . They did not rise to the opportunity the way I'd hoped (author's interview).

This lack of influence by one of the few candidates for membership in a scientific ecology epistemic community suggests that such a community, if it existed at all, had a limited substantive impact at UNCED, and little influence on governance norms or the framing of issues more generally.

The discussion so far should not suggest that scientific evidence was unimportant to UNCED, but rather that it was not the driving force behind norm creation or the initiator of action. The broad shape of the norm-complex articulated in the Rio Declaration and Agenda 21 does not reflect the primacy of "scientific ecology" as the basis for agreement, and the Declaration in particular appears less concerned with environmental protection or ecological concerns than even the Stockholm Declaration. Agenda 21, while it incorporates insights and linkages identified by scientists, also reflects the environment and development mix of liberal environmentalism that did not come primarily from science. As at Stockholm, scientists achieved the most success on their own turf. However, much of the science used at UNCED came from governments and the secretariat-commissioned reports, not from the independent influence of an epistemic community. Scientists were not particularly active outside of those limited roles.

Science and Climate Change

To be fair, one cannot measure the influence of an epistemic community by looking at conference preparations alone. Indeed, the power of epistemic communities can occur in their ability to frame the issues for negotiations rather than changing negotiation outcomes *per se* (although the latter position is often taken in empirical tests of the literature, where the community is credited with forming specific focal points for agreement). Hence, below I examine in some detail the influence of the scientific community in the lead-up and follow-through of international action on climate change that led to the signing at UNCED of the Framework Convention on Climate Change (FCCC). Space limitations prevent me from undertaking similar detailed analyses for biodiversity and forests agreements.

Admittedly, the biodiversity case does provide some support for an epistemic communities hypothesis since the long-standing activities of scientists affiliated with organizations such as the IUCN and later UNEP did play an important role in promoting the biodiversity concept. Scientists also helped to define issues and propel bargaining to produce international action around the concept.[38] However, even on biodiversity, the area closest to traditional conservationist concerns, ideas that shaped the overall agreement included the range of norms discussed in chapter 3 that had little to do with the relevant science. Major debates in negotiations revolved around intellectual property rights and sovereign control—debates that shaped the core normative basis of the treaty, making it a good fit with the broad normative contours of liberal environmentalism.

Climate change is an appropriate focus for a number of reasons. First, most analysts agree that climate change, although not even officially a part of the UNCED process, became a main galvanizing issue for action, and mirrored many core debates (Imber 1994; Boehmer-Christiansen 1994c: 181). It was also a central outcome of the UNCED process. As such, many of the norms found in FCCC mirror those found in other UNCED documents negotiated at the same time. Second, climate change, especially as part of the broader agenda of global change, epitomizes a problem appropriately framed in ecological terms. Its very definition implies complexity, interaction of various environmental media (land, sea, and air and the chemical, physical, and biological cycles that link them) and their relationship to human activity (anthropogenic change), and for solutions to take account of those interactions. It also represents, perhaps more than any other problem except ozone depletion, a truly global issue. In fact, climate change is often used interchangeably in practice with the umbrella research program of "global change," which by the 1990s had become the most prominent global environmental research program. States, international organizations, and non-state actors have devoted an enormous amount of effort and resources to this issue, which, owing to its potential implications and a scope that encompasses a wide variety of other environmental and non-environmental concerns, has become the dominant environmental issue on the international agenda. Finally, although consensus appeared uncertain at times, the scientific community around climate change was well organized and mobilized in terms of promoting international action. Here I focus not only on how science worked in the negotiation and treaty process, but also on how much it influenced the form the treaty finally took, hence how it helped shape global environmental governance more broadly. The story is told more or less historically to show how scientific ideas about climate change made their way onto the international agenda.

The climate change case does show that scientists can affect international action on a highly technical issue. However, typical of other examples and the broad evolution of governance explored above, that influence did not work in the way the epistemic communities hypothesis suggests. There was no consensus on values of "scientific ecology," early success in getting international attention did not translate into control over how the problem would (or would not) be addressed, and science eventually got molded by the political process and normative structure as much as or more than it molded them.

Fears of human-induced climate change are nothing new. Since the mid-eighteenth century, scientists have arguably used this threat as one of the few effective instruments to persuade governments of the seriousness of environmental change. The so-called greenhouse effect (that naturally occurring carbon dioxide and water vapor keep the Earth's temperature about 33 degrees C higher than it would otherwise be) has been known since the nineteenth century. The two concerns became linked when in 1938 G.S. Callender found higher concentrations of carbon dioxide in the Earth's atmosphere than in the nineteenth century and that human burning of fossil fuels since the industrial revolution began could account for the differences. He also suggested that global warming might result, although his findings were greeted with much skepticism.[39]

It took another thirty years for a sustained transnational research program, which began with the Global Atmospheric Research Programme (GARP) in the mid-1960s, a collaborative effort of ICSU and the WMO. UNESCO's Man and Biosphere Programme also researched the issue, which often overlapped with ozone research as the problems are related. However, political activity around ozone, until at least the late 1980s, generally ignored the links present in major scientific studies. An epistemic community of sorts could be traced to these efforts as key personalities would later become involved in GARP's successor, the World Climate Research Programme (WCRP), ICSU's International Geosphere-Biosphere Programme (IGBP), and the Intergovernmental Panel on Climate Change (IPCC). A prominent example is Bert Bolin, who from 1988–1996 headed the IPCC, mandated since 1988 to advise governments, UNEP and WMO (its sponsors) and the United Nations system as a whole on climate policy. Governments also mandated IPCC to update, transmit and assess relevant scientific information and point out policy implications. The overlap in IPCC and global research bodies at senior levels is well documented.[40] In addition to climate scientists, the core group of researchers in these organizations and at IIASA included energy demand forecasters.[41]

Yet again, IIASA deserves special attention for its long-standing involvement in the issue, although its direct influence on climate change policy is difficult to discern. However, Boehmer-Christiansen (1994a:146) suggests that the potential growth of influence of IIASA, especially through nongovernmental groups, "possibly became threatening to some governments and governmental science, encouraging efforts to capture climate change research from the private sector," a task at which governments proved quite successful, as I explain below. IIASA was one of the few organizations that explicitly linked climate change to "sustainability," broadly defined, and key members of the climate research community had connections to IIASA. For example, William Clark, who headed up the sustainability project, also delivered a keynote address on policy at the Villach conference in 1985 that helped spur the scientific community into political action on the climate change issue.[42]

The meeting in Villach, Austria marked the turning point toward a sustained transnational scientific research program aimed at generating international political attention. Although the first World Climate Conference had addressed the issue in 1979, the Villach conference marked the real beginning of efforts to build a scientific consensus.[43] The relatively late start to consensus building was not owing to a lack of attention by scientists. On the contrary, since the mid-1970s, IIASA, the International Energy Agency, and even the OECD had held high-profile conferences and raised concerns about climate change. Around the time of the first climate conference, UNEP began to get involved and it was UNEP—not WMO, which had most of the governmental science expertise on the issue—that attempted to link climate change to development policy. UNEP also encouraged SCOPE to consider such linkages in its research on the carbon cycle. UNEP head Mustafa Tolba, a botanist by training, encouraged this trend, but his concerns appeared to stem primarily from his political convictions and concern for the developing world, not his scientific credentials or background. He had hard work ahead of him to bring scientific research around to this point of view (Boehmer-Christiansen 1994a:155).

Villach pulled together high-profile governmental and nongovernmental scientists and what might be called research brokers or science managers.[44] Jointly sponsored by UNEP, ICSU, and the WMO, this conference represented the core of an epistemic community on the climate change issue. It also explicitly aimed to influence policymakers, a position in line with the general philosophy of all three sponsoring organizations. Other high-profile transnational scientific organizations gave institutional support, notably IIASA and the Beijer Institute (which later became the Stockholm Environment Institute). Also present were high-profile U.S. environmental organizations such as the Environmental Defense Fund and World Resources Institute, and national research institutes such as two Max Planck Institutes in Germany.[45] James Bruce, an Environment Canada scientist and assistant deputy minister at the time, chaired the conference.

Villach came out with a strong, unified position that an unprecedented increase in global mean temperature could occur in the first half of this century. Current trends in greenhouse gas (GHG) emissions (including gasses other than CO_2) would lead to the equivalent of a doubling of CO_2 in the atmosphere during this time period. If left unchecked, GHG concentrations could lead to an unprecedented, and potentially catastrophic, increase in global mean temperature from 1.5 degrees C to 4.5 degrees C.[46] A follow-up workshop in Bellagio in November 1987 recommended that science-based targets should be designed to limit temperature increases to 0.1 degrees C per decade. However, there is arguably little "scientific" basis of this 0.1 percent limit, especially since what is socially tolerable or dangerous involves value judgments (Agrawala 1999:164). These early prescriptions came out much stronger than the IPCC report of 1990 on the policy side, although more cautious on the science. Thus their main emphasis was the high level of uncertainty and need for greater research. Apart from climate science, research and development was to focus on alternative energy technologies and policies. Significantly, the one nonconsensual document at Villach was Tolba's "agenda of action," which would have made UNEP the main center for policy, while others present thought organizations such as IIASA were better placed to give policy advice.

Participants at Villach set up the independent Advisory Group on Greenhouse Gases (AGGG) in July 1986, under the auspices of WMO, UNEP, and ICSU. F. Kenneth Hare, a respected climatologist from Canada, who had also helped organize the 1979 First World Climate Conference sponsored by WMO, chaired this small group of experts at the very top of their fields. All were prominent in various transnational research efforts or organizations and individually their work had laid much of the basis for current climate change research. Other group members are as follows, listed with just some of their credentials: Bert Bolin (later the first chair of the IPCC); Gilbert White and Mohammad Kassas (both worked on water resource management and land degradation issues, and were associated with ICSU and UNEP); Syukuro Manabe (a pioneer in general circulation models and the leading modeler at the Geophysics Fluid Dynamic Lab at Princeton), Gordon Goodman (head of Stockholm's Beijer Institute, who focused especially on energy policy issues at this time), and Gueorgui Golitsyn (who headed the Institute of Atmospheric Physics at the Soviet Academy of Sciences). Other younger researchers and "advocacy scientists" coalesced around the Beijer Institute under Goodman, and undertook activities sanctioned by AGGG.[47]

Organizing institutions mandated the AGGG to monitor climate research data, conduct assessments of increases in GHG concentrations and effects, advise governments on possible mitigation measures, and possibly to initiate consideration of a global climate convention. Along with other organizations, AGGG followed up on Villach with a second workshop there and another in Bellagio, Italy (mentioned above) in 1987. These conferences are notable for increasing calls for political action and greater participation by policymakers. AGGG clearly constituted an epistemic community concerned about the potential environmental impacts of human-induced climate change, although it also included energy-demand forecasters. Agrawala describes the early days of the AGGG as "an almost utopian era where a small network of experts, international organizations and environmental advocacy groups had a near monopoly both on the international science and policy agenda" (1999:158).

The AGGG marked the high water point of scientific leadership on the climate issue, but that "utopian" era was short lived, as the IPCC gradually overtook the AGGG's work and place in global climate policy. Its influence culminated with the June 1988 Toronto Conference on "The Changing Atmosphere: Implications for Global Security," which grew directly out of recommendations and findings of the Bellagio workshop.[48] The Toronto conference director, Howard Ferguson of Environment Canada, who had attended Bellagio, worked closely with Canadian Environment Minister Tom McMillan to marshal the Villach and Bellagio findings to produce a strong policy statement, which McMillan wanted. Ferguson also recruited AGGG members to be on the steering committee. Among other members, the committee also included Jim MacNeill, who had just finished his work on the Brundtland report, another impetus for the conference. The presence of Gro Harlem Brundtland, Canada's Prime Minister Brian Mulroney, and number of ministers from a G-7 Summit held earlier at the same venue, among the 300 scientists and policymakers present, gave climate change science its most influential audience to that point.

The main conference recommendation, that governments and industry should reduce CO_2 emissions by 20 percent from 1988 levels by 2005 "as an initial global goal" was the most powerful and specific policy recommendation to that date, and became a rallying point for global action in early climate change negotiations. The conference statement also recommended the development of "a comprehensive global convention as a framework for protocols on the protection of the Atmosphere."

These high-profile and credible policy conferences also happened to correspond with a series of external events and political and economic circumstances that, although largely unrelated to science, elevated public concern and galvanized government responses to climate change. Economic factors included falling fossil fuel prices and growing fuel and energy technology competition, which gave alternative energy suppliers (that is, not coal and oil) incentive to support climate change research and action. Political factors included the winding down of Law of the Sea and acid rain negotiations that had preoccupied countries interested in environmental negotiations. Most of all, high-profile environmental events such as the discovery of the ozone "hole," the Chernobyl disaster, and especially the unusually hot summer, and drought, in North America in 1988 raised public concern about global environmental problems.

Heightened media attention culminated with the famous U.S. Senate energy committee testimony of NASA scientist James Hansen, who said research supported a causal relationship between an increased greenhouse effect and observed higher temperatures over decadal time scales, which could cause increased heat waves and droughts. With temperatures reaching a record 101 degrees F., Hansen, chief of NASA's Goddard Institute of Space Studies and a climate modeller, told the hearing "the greenhouse effect has been detected and it is changing our climate now," and suggested that politicians should "stop waffling" and take action (Paterson 1996:33; Boyle 1999). It would, however, take another seven years for scientific consensus, when the 1995 IPCC Second Assessment Report found that, "The balance of evidence, from changes in global mean surface temperature and from changes in geographical, seasonal and vertical patterns of atmospheric temperature suggests a discernible human influence on global climate" (IPCC 1995:2.4).

The summer of 1988 marked the pinnacle of epistemic community influence. Until then, most Western governments viewed climate change as mostly a scientific and environmental problem (Bodansky 1994:50). The United States, however, from the start viewed the issue from a more economic perspective and through the lens of domestic policy. Thus, while other states primarily dealt with the issue through environment ministries, the U.S. administration set up its own committee of the White House Domestic Policy Council, having learned from ozone negotiations that the EPA and State Department might move more quickly than the White House desired. Although EPA had representation, the major players included the powerful departments of Energy, Commerce, and Interior, the Office of Management and Budget, and the Council of Economic Advisors (Bodansky 1994:50; Victor 1995:365). The result was a policy position that emphasized measuring economic costs and cost/benefit calculations of environmental risk. The combination of domestic interests including a powerful fossil fuel lobby and a conservative White House who viewed environmental problems primarily through an economic lens meant scientists were essentially outmuscled in U.S. policymaking.

Internationally, pressure from the United States especially, but also other industrialized countries, led to the marginalization of the work of the AGGG as climate change turned from a primarily scientific to a political issue. James Bruce, secretary of the WMO Executive Council when it decided to set up IPCC, described the shift from the AGGG to the IPCC this way:

> ... after a couple of years of their [AGGG] work, there was an unease ... that crept into some governments that this was an issue that was going to have enormous economic repercussions one way or another and they, in particular the United States, didn't like the idea of these free-wheeling scientists pronouncing on the subject. They preferred something with more governmental involvement (author's interview).

This unease about "the sort of influence that a semi-independent group of scientists might have," as Bruce put it, led the WMO and then UNEP to jointly sponsor the *Intergovernmental* Panel on Climate Change, which, as its name suggests, was more directly under the control of governments. The idea of an IPCC had percolated within WMO since at least May 1987, overlapping with the AGGG, and AGGG work continued until 1990, but for all intents and purposes the IPCC took over the policy role of AGGG upon its launch in November 1988 (Agrawala 1999). In addition, the original members of the AGGG had already begun to drop out of the group. Some expressed increasing discomfort with their policy role because they worried that advocating policy as science, would compromise their credibility (Agrawala 1999:164, referring to a published interview with Syukuro Manabe).

In December 1990, governments also took the actual negotiations toward a convention out of the hands of the WMO and UNEP (unlike ozone, for example, which was negotiated under the auspices of UNEP) and put them into the hands of the UN General Assembly. Under Resolution 45/212 on Protection of the Global Climate for Present and Future Generations the UNGA set up the Intergovernmental Negotiating Committee (INC), handing over negotiations fully to diplomats and out of the hands of the IPCC, which still operated independently of the negotiations themselves.

Since the FCCC came into force in 1994, new subsidiary bodies that institutionalized the role of scientists further entrenched government control (biodiversity and other agreements contain similar bodies), although the

IPCC remains as a source of advice "independent" of the convention. Governments approve the scientists who sit on subsidiary bodies responsible for processes such as periodic reviews, evaluations of triggers for further action, monitoring, and joint research activities. Despite the importance of such functions, no guarantee exists that parties will accept the findings of such groups or abide by their recommendations since political pressure at home, for example, might lead them to usher counter scientific evidence, or use political or economic arguments to suggest alternative actions, no action, or the need for more research (Susskind 1994:65). The increasing level of government control also means such bodies are less likely to embark on independent initiatives in terms of shaping the overall governance structures, which have essentially been set for them. Actual allocative decisions or objectives, then, are not likely to stem from these bodies, though their findings may be used to suggest the need for swifter action, for example.[49]

The combination of these events served to compromise the independence of the climate change research community, which suggests it can no longer be considered an epistemic community according to the definition given earlier since members now consist of government-approved scientists. Bruce, although he defends the IPCC's scientific integrity, largely confirms how the role of science changed as the political stakes grew. One such change involved the replacement of natural scientists largely by economists in formulating policy options.[50] Nonetheless, the scientific community sees the IPCC as its main voice in politics and policy.

The first two IPCC reports (1990 and 1995) also present a window through which to see the politics of scientific activity on climate change. The IPCC divided into three working groups. For the first assessment report, working group I provided the scientific assessment of climate change, working group II the impacts of climate change, and working group III response strategies. In the negotiations that led up to the climate change convention, the working group I report received the most attention, although its influence apart from warning about the problem was limited. It was also somewhat divided over its emission scenarios, the part of the report most likely to affect policy because projections on emissions are a necessary part of determining what kind of action would be required to prevent "dangerous" levels of increases in greenhouse gases (the objective of FCCC as found in article 2) (Boehmer-Christiansen 1994a:148–149).

Other significant recommendations came from working groups II and III on the need for more research into the sensitivity of "socioeconomic" systems to climate change. Working group III also noted the lack of cost-benefit analysis or research on technological or market impacts of proposed solutions. The findings had little impact on the treaty process, as working group III was largely ignored by policymakers, but findings proved significant for the future research agenda of IPCC.

As a result of the report, and pressure especially from the United States, which favored cost-benefit analysis of environmental problems and was moving to support market-based solutions, working group III was completely recast in 1992 to undertake the second assessment.[51] Its new mandate focused much more on economic modeling, specifically to conduct "technical assessments of the socioeconomics of impacts, adaptation, and mitigation of climate change" (IPCC 1995, Working Group III Summary for Policymakers:1). Whereas the original policy group consisted of a mix of scientists, engineers, and administrators, economists dominated the new group, reflecting also the growth in the academic field of the economics of climate change. It also represented a shift in emphasis from technical solutions and opportunities favored by countries such as Japan, to the economic costs and benefits of various responses and the policy instruments to best achieve them.[52] Other social scientists (political scientists, sociologists, geographers, and so on) were not generally selected by governments, hence the report has little to say about sociopolitical factors such as societal stresses, changes in government, institutional adaptation, and so on.

Although it has been the object of some controversy over subjects such as differential costing of "statistical" lives in developed and developing countries, the 1995 findings of working group III have generally received more attention by policymakers than did the 1990 report.[53] This increased impact seems likely a result of the report's more economistic rather than ecological approach to policy.

The motivation of lead researchers on climate change, a main component of the epistemic communities hypothesis, is difficult to determine in aggregate. Interviews and documentary evidence suggest that many shared causal beliefs about the nature of global change, but consensus on principled beliefs or a desire to "develop social laws from their understanding of the laws of nature" finds little support.

For example, Boehmer-Christiansen, in her extensive study of the climate change policy process, argues that the coordinated research community "acted primarily as a lobby for its own research agendas dedicated to the modeling of planet Earth and the development of alternative energy sources" (1994a:140; 1994b). Expensive technologies for modeling, the performance of which in large part drove the success of the climate research in the mid-1980s, perpetuated this need. Other research programs related to energy forecasting, for example, also used climate change to convince governments of policy relevance at a time when fossil fuel prices were dropping. Boehmer-Christiansen argues that scientific bureaucracies used climate change to convince governments of policy relevance and the need for further research at least in part as a way to gain public money (even at the expense of increased government control) to fund these programs beyond the means of the private sector. Most other literature takes a less cynical view of motives, but emphasizes only that scientists pushed for some policy response, not that they tried to develop policy norms based on "social laws" derived from nature.

Evidence also suggests that splits existed in the scientific community on what scientific findings meant in terms of policy implications. Moreover, many scientists, even the most active, such as former IPCC chair Bert Bolin (1994), have expressed reluctance to enter debates that speak too directly to actual policy choices. Admittedly, a coordinated research community certainly emerged after 1985 that shared a consensus on the nature of the problem and agreed that it ought to be brought to the attention of policymakers (although the group also acknowledged a high degree of uncertainty). However, there was and remains much less consensus on the principled beliefs about how to think about the threat from climate change in terms of resiliency of ecosystems and the like (James Bruce, author's interview). If one principle did arise, it was the notion of precaution in the face of uncertainty.

In terms of an ecological approach, even the most ardent supporters of the values of an "ecological scientific community" did not press hard for that approach in policy. For example, the initial approach, advocated by Canada, called for a framework agreement on a "law of the atmosphere," like the Law of the Sea. Agreement on a framework convention could then lead to separate protocols on specific issues such as acid rain, ozone depletion, and climate change. "The rationale for this approach was that it recognized the interdependence of global atmospheric problems" (Bodansky 1994:53). Similarly, the Toronto Conference Statement called for "a comprehensive global convention as a framework for protocols on the protection of the atmosphere," which fit well with a "scientific ecology" orientation. Ironically, a second approach later adopted—to focus simply on a convention on climate change—came from Tolba, whose stature was high because of his leadership role in producing an international agreement on the ozone issue. At a conference in Ottawa in 1989, Tolba strongly criticized the "law of the atmosphere" approach as politically unrealistic, and argued for a more narrowly focused convention. As a result, it never again achieved serious consideration (Bodansky 1994:53). So it would seem that when it came to policy, the scientists most actively engaged seemed as driven by political expediency as by drawing social laws based on the laws of nature.

The actual content of the climate change treaty, while it certainly does reflect some ecological ideas, also has embedded in it the core norm-complex of liberal environmentalism. For example, the convention's objectives include (in Article 2) stabilization of greenhouse gas concentrations "at a level that would prevent dangerous [not defined] anthropogenic interference with the climate system . . . achieved within a time frame sufficient to allow ecosystems to adapt naturally." However, it goes on to say that the level decided upon should also "enable economic development to proceed in a sustainable manner."[54] The principles and commitments reflect this balancing act, reproducing the same or similar wording to principles found in the Rio Declaration. Scientists certainly did not have consensus on this set of norms, nor did they clearly articulate a set of values that flowed from scientific research that would frame the convention differently.

In addition, virtually no discussion occurred prior to the Toronto meeting about North-South issues. The scientists who dominated atmospheric research primarily came from the North and this group, for the most part, did not address discussions about the effects of climate change on development. The only notable exception came out of the second (1987) Villach workshop where delegates recognized that aid might be necessary to pay for anticipatory adaptation to climate change in developing countries (Rowlands 1995a:189). Scientists simply did not deal with these issues, which they perceived as political.

Even after the treaty was signed in 1992, entrenching ideas inherent in the concept of sustainable development, scientists did not much discuss the concept, but simply accepted it as part of their mandate. The understanding of sustainable development within working group III of the second assessment report was a basic notion that severe environmental or economic damage would make development unsustainable. As Bruce, co-chair of the working group, put it, "If [the population is] going to be flooded, then that makes it difficult to sustain development for a small island state . . . or if [the state is] going to have economic losses of nine per cent GDP per year" (author's interview). He added that the ecological perspective did not dominate thinking about sustainable development among scientists. Moreover, they often viewed issues around sustainable development as political questions not fit for recommendations by the IPCC. For example, the first IPCC report largely left equity issues unexplored and the second report took a cautious approach, emphasizing that politicians should choose between such policies although the effects could often be determined scientifically.

It is not therefore surprising that the IPCC itself simply had little direct effect on the content of the FCCC signed at Rio. A commentary on the 1990 IPCC assessment report by authors affiliated with IIASA put it as follows:

> The first report [on the science of climate change] is easily the single most referenced document on the science of climate change, demonstrating its focal role in the public debate. IPCC "consensus" documents are paraded by both environmental non-governmental organizations (ENGOs) and governments in their domestic debates to "show" that the climate problem is real and deserves policy action—and by others to demonstrate the reverse—even though the IPCC documents are appropriately more cautious. . . . [However], in practice, it has been extremely difficult to integrate research on the effects of climate change and policy options into coherent and useful consensus documents (Victor and Salt 1994:27).

They note elsewhere that a reorganization for the second assessment report explicitly included more economics and expertise drawn from developing countries, two developments driven much more by governments than the scientific community or the science of climate change (Victor and Salt 1995). Rowlands (1995a:89) is more blunt, stating that the 1990 IPCC report was marginalized in the intergovernmental negotiating process and that "politicians regarded any consensual scientific knowledge as but one of many inputs."

The second assessment report (1995) responded to these concerns on the policy side, but its presentation of options (its mandate) does not demonstrate a consensus. Furthermore, the themes that do emerge draw much more on economics than they do on a vision typical of ecological science. This result is not surprising, given that social scientists who dominated working group III were primarily economists. Most of the economists approved by governments, although certainly prominent, were classical and/or environmental economists, not ecological economists. In fact, hard scientists and economists often disagreed on policy instruments and approaches (and there were debates on technical issues among economists as well) (James Bruce, author's interview). While the overall approach reflects the mandate given to the IPCC (and taken from the convention) to "place the socioeconomic perspectives in the context of sustainable development," it carved out its analysis more narrowly to reflect liberal environmentalism.

This disjuncture between research programs of natural scientists and social scientists persisted within research circles throughout the work of the IPCC. Natural scientists continued to approach the problem of climate change in terms of physical flows of matter and energy, while social scientists defined environmental problems in terms of human behavior. Cohen et al. (1998) illustrate the conflict between climate change framed as a scientific problem and the discourse of sustainable development, which they see reflected in the two most prominent transnational research programs, the IGBP and the International Human Dimensions Program (IDHP), dominated by social scientists. They note a "puzzling lack of interaction between two fields of research and activity, associated, respectively, with climate change and with sustainable development" (1998:342). Policymakers were relatively free to shape the social scientific analysis as well, along lines already suggested within a narrowly economistic framework. Although this made the work more policy relevant, it tended to be weak on broader sustainable development concerns. As Cohen et al. (1998:342) note, although the mandate of working group III of the second assessment report was to address sustainable development issues, "the actual discussion of SD [sustainable development] is almost non-existent." This limitation of the research might explain why policymakers could relatively easily treat scientific findings as less policy relevant, except in the minimalist sense (in policy terms) of identifying the environmental problem and biogeophysical impacts (rather than social impacts or policy responses).

Three conclusions on the climate change case raise difficulties for the epistemic communities approach to understanding international environmental governance. Recall, climate change should be an easy case, which suggests the challenges to the hypothesis ought to be taken seriously.

First, scientists could not or would not formulate a coherent set of policy ideas that states then adopted. Admittedly, a transnational scientific community did prompt an international policy response independently and ahead of public concern with the climate change issue. In that way, the epistemic community hypothesis finds some support in the climate change case. However, scientists did not have a large influence on policy formation. The height of influence probably came in 1988 with the Toronto conference target of 20 percent cuts. Although environmental NGOs and states likely to be most affected by climate change (for example, small island states) often raised this concrete recommendation prior to Kyoto, it never received serious consideration as a focal point for agreement. In addition, the ecological approach did not dominate proposals. Governing norms that stemmed from the scientific community, apart from the Precautionary

Principle, either did not appear in the agreement reached or did so in the context of other norms that supported liberal environmentalism. The one policy implication, apart from reducing emissions, that seemed to produce consensus in the scientific and technical communities involved in climate change research was the need for policy to focus on alternative energy technologies and policies. But that concern too gradually eroded as a focus for policy, even within the IPCC, which concentrated more on the economic efficiency of policy options.

Second, when scientists did gain public/political attention, they did so through encouraging concern on the issue itself, not its framing. In any case, governments quickly responded to such concern by taking control of transnational research and policymaking. Governments also took allocative decisions out of the hands of scientists, who proffered few unified recommendations anyway on such issues. Far greater consensus could be found among economists who seem to be increasingly important providers of policy advice.

Third, the linkage of scientific activity and ideas to sustainable development came late, and was largely pulled by individuals with existing links to the development community, such as UNEP head Mustafa Tolba. These links did not spring from science or the climate change research community. In Tolba's case, the concerns stemmed from his official position and his own convictions. In numerous published speeches he stressed the importance that should be placed on the effects of climate change, among other environmental problems, on the developing world. His speeches also emphasized the need to formulate problems in a way to address such concerns. In contrast, most of the relevant scientific community demonstrated either indifference to sustainable development as framed by UNEP or unease with UNEP playing a leading role (Gordon McBean, author's interview). The conclusions on climate change are indicative of the overall performance of the epistemic communities hypothesis in explaining the evolution of international environmental governance.

CONCLUSIONS

This chapter has questioned the proposition that science is a primary informer of policy direction on international environmental concerns. Thus it challenges a key conclusion of Peter Haas's, "that science is essential for the understanding of global environmental problems, thus shifting the determination of the scope of allocative decisions to the international institutions for science" (1996:1). To the contrary, scientists were largely excluded from allocative decisionmaking and often eschewed such roles. When they did have influence, such as in promoting the Precautionary Principle, it did not come as a direct outcome of their specialized knowledge of cause-effect relationships.

Second, consensus on cause-effect relationships within scientific communities did not seem to correlate well with action on major issues, although sometimes individuals or groups of scientists played an active role in promoting particular environmental concerns. On the central question of principled beliefs, consensus often seemed particularly weak on a number of dimensions, making the case difficult to sustain that such consensus was either necessary or sufficient for development of particular norms of environmental governance. Even on specific matters of policy choice, consensus was relatively uncommon. Thus the consensual knowledge which supposedly gave a potential epistemic community its political power is open to question.

Third, problems arise from the literature's emphasis on tracing the effects of single communities, based on the assumption that a particular group should be privileged because of its claim to authoritative knowledge in the particular issue area. This approach can easily miss the competition of expert groups who come at policies with different agendas. Environmental policy since 1972 has not been the sole concern of a community of scientific ecologists. As the development of norms in practice suggests, development and environmental economists also had an interest in environmental policy and actively sought, or were solicited, to influence the shape international environmental governance would take (a point taken up in the next chapter).

A related problem is the focus in the literature on an ideal-type scientific ecology epistemic community. This focus has led Haas to the erroneous conclusion that environmental governance now faces a backlash from rules and principles of trade regimes and market challenges at the domestic level (Haas 1996:43–44). This bias is built into epistemic community studies since they start with the ideas of a particular community and assume a backlash when those ideas do not dominate. The bias closes off a critical examination of how such ideas interacted with other forces or whether they are indeed the basis for the norms and institutional arrangements that finally result. Hence the focus on "scientific ecology" simply misses the compromise of liberal economic and environmental norms that was at the very center of bringing environmental norms into mainstream international governance. Although arguably the compromise embodies important contradictions, it has shaped the way environmental concerns are now institutionalized in international governance and arguably it has been the single most significant factor in shaping international environmental governance over the last thirty years. The so-called backlash does not exist; it is instead a logical outgrowth of the norm-complex developed over that time period.

A second set of conclusions concerns how science actually did work. The primary pattern revealed is that, contrary to the epistemic community hypothesis, scientists were reactive, not proactive in the major norm-articulating events identified, even learning themselves from their involvement in such activities.

The science policy literature often focuses on reasons for a lack of independent influence, including mutual distrust of scientists and policymakers/publics or simply a reluctance of scientists to enter into the policy process, even among those most active in communicating scientific ideas to policymakers.[55] Similarly, Lemons and Brown (1995:13) note the "fallacy of unfinished business"—the tendency for scientists to see problems as technical, therefore requiring only technical solutions. That view closes off for them social, ethical, or political solutions. Poor interdisciplinary communication may also limit the production of policy-relevant knowledge.[56] Such problems may also stem from the difficulties of effective communication between scientists and policymakers and the public, which leads to misunderstandings especially in the face of incomplete knowledge and scientific uncertainty (Rowland 1993). Often policymakers can then choose among competing knowledge claims within scientific communities.[57]

It should not be surprising, then, that scientists involved in international environmental policy exhibited the same kind of unease as their colleagues in other policy-relevant fields when they got too close to political processes. This situation remains largely unchanged since 1972. At that time, Lynton Caldwell (1990:115) noted that SCOPE "appears to accept an assumption widely shared among scientists who believe that their public mission is largely fulfilled when scientific studies are made available to governments and international organizations."

William Clark similarly looked back on the Brundtland Commission process and argued the small role science played was for the best:

I think it's probably just as well. . . . The agenda on sustainable development moved ahead very rapidly in the Brundtland era into UNCED in ways that were largely successful in shaping a political re-framing of the terms of the debate, a political consensus on at least some directions we

needed to be heading. . . . Frankly, they had about enough science to let
that go forward and not so much that it got in the way. [Now it is time]
for the science community to make a re-examination of the sustainability issue and see whether, given the political consensus that has shaped
up . . . we're doing the right science . . . whether the necessary long-term
science and monitoring legs for this venture fit . . . the development, the
political, and economic legs. So I think [the science and politics of sustainable development] were out of phase, but whether that was done in-
tentionally or not, [it was] probably done very effectively and would
have been less effective had the scientists been running all over the
Brundtland Commission (author's interview).

Such a conclusion suggests the causal arrow runs opposite to the way the
epistemic communities hypothesis suggests. Even when groups of scientists
attempted to maintain their independence, governments proved particularly adept at reining in science and setting parameters for their research and
influence on policy. Interestingly, in one of the few detailed comparative
studies of the science-policy nexus in major international environmental
agreements, Andresen et al. (2000) find that scientific influence on policy,
although weakly correlated with involvement in the policy process, may be
inversely related to autonomy of scientists in the policy process (Underdal
2000b:196–199). While not completely contradictory to an epistemic communities emphasis on the need for scientists to infiltrate domestic bureaucracies and international institutions, it does suggest at least a healthy skepticism that the influence of epistemic communities will reflect their own
"worldview" once involved closely with the policy process, since governments and other political factors influence the research process to their own
ends. Underdal argues that this finding suggests that scientists may perform
different functions at different stages of the policy process, or in different
institutional settings. Also consistent with the findings here, Andresen et al.
found that while scientific evidence "often serves to precipitate *some* kind of
policy response" and can play a role in agenda-setting, "The *substance* of
that response, however, is determined essentially by politics rather than science" (Underdal 2000b:184, emphasis in original).

Finally, a number of empirical anomalies suggest that the evidence simply does not hold up in defense of a narrow focus on science in explaining
the evolution of environmental governance. Most obviously, there is a lack
of fit between ideas generated by an epistemic community of natural scientists and observed normative shifts. The uninterest with which most scientists greeted the political/economic questions that seemed to guide much

of norm creation suggests that a key condition for such an epistemic community's influence has not been met.

One broader normative concern about the epistemic community literature deserves mention before closing off this discussion—that the literature makes an assumption that if consensual knowledge exists on environmental management in an epistemic community, then institutions ought simply to be designed to better integrate such knowledge. That position ignores the possibility that contestation might come from outside that community, or that the community is not equipped to deal with the broader social and political implications to which that knowledge might be put to use. Conversely, that position might blind analysts into assuming that the epistemic community is being listened to when the problem it identifies is being addressed. Such an assumption makes it easy to ignore the real contestation over how the problem is being addressed, to what ends, and for whose benefit. Thus, epistemic community analysis either misses the boat on how ideas inform governance, or leads to an uncritically examined normative end point where the community's prescriptions are assumed to be in the best interest of humankind.

ECONOMIC IDEAS, SOCIAL STRUCTURE, AND THE EVOLUTION OF INTERNATIONAL ENVIRONMENTAL GOVERNANCE

THIS CHAPTER MAKES two basic theoretical points. First, it argues that ideas that become institutionalized as governing norms must find some "fitness" with the existing international social structure. This is true because most issue areas that constitute coordination or collaboration problems for states generally exist within a nested set of governing norms that have legitimacy. Second, this chapter argues that new norms may come from a variety of sources, but that these sources must have a basis of legitimacy themselves in the eyes of key actors who participate in, and are affected by, the governing structures they create. The key to understanding the evolution of international governance, then, is to try to gain an understanding of the interaction of new ideas with the social structures—institutionalized or nested sets of norms—they encounter.

Substantively, this chapter argues that economic ideas overshadowed scientific ideas and ecological thought in producing normative compromises at key junctures over the last thirty years of international environmental governance. In light of the two theoretical points just made, that influence was not accidental. Policy entrepreneurs[1] were most successful at moving a concern for the environment into the mainstream of international governance when they tried to nest environmental norms into the broader international social structure, even as that structure evolved to reflect the now dominant liberal economic order. The result was the compromise of liberal environmentalism. Actors and the ideas themselves certainly mattered. However, the success or failure of new ideas in response to environmental problems must be understood in the context of their interaction with the wider international social structure they encountered and attempted to modify.

Looking back to the origins of international environmentalism, it may seem that a single line of economic thinking inevitably led to liberal environmentalism today. Indeed, many ideas associated with liberal environmentalism received some policy attention as far back as the late 1960s. Their roots go back even further to Pigou's book *The Economics of Welfare*, which suggested that governments should introduce corrective taxes and subsidies to discourage activities that generate externalities.[2] The basic elements of Pigovian analysis became the foundation for the new field of environmental economics that focused on ways to internalize the ostensibly external costs of environmental degradation.

The late 1960s saw a revitalization of these ideas in the academic literature. For example, in 1968, University of Toronto economist John Harkness Dales's book *Pollution, Property and Prices* introduced the idea that transferable property rights could work to promote environmental protection at lower aggregate cost than conventional standards (Dales 1968; Thompson 1972). By the early 1970s, a number of government and academic studies that contained specific suggestions on how to cost the environment had appeared in the United States, the United Kingdom, and Canada.[3] The ideas promoted in those studies ranged from the development of economic incentives to the creation of private property in pollution rights, based on Dales' ideas. As Hahn and Stavins (1992:464) note, "From these two seminal ideas—corrective taxes and transferable property rights—a substantial body of research has developed." That research would eventually have a profound effect on environmental policy.

However, other lines of thought about how to bring economic activity more in line with environmental control had also appeared during this period. The policy literature in the lead-up to the 1972 Stockholm conference contains a number of alternatives that draw on different traditions in economic thinking. For example, some policy proposals advocated an extreme version of command and control, which suited a position on environmental problems taken by eco-pessimists. Proponents of this view saw environmentalism as eschewing the left/right, or East/West ideological divide since environmentalists in the West would realize that private property as it then existed would no longer be possible for natural resources. For example, one writer in 1972 noted that "in the United States, for example, one may read articles affirming that one may one day come to the point where all natural resources will have to be state-controlled, if not declared the property of the community altogether" (Caponera 1972:139). During this period, common property ownership and community rights to enforce standards applied to that property were as seriously considered as private ownership schemes.

More recently, Eric Helleiner (1996) has identified a distinguishable "green" perspective on International Political Economy (IPE). This perspective, Helleiner argues, differs in its normative goals and theoretical assumptions from the traditional realist, liberal, and Marxist variants of IPE. In essence, the "green" perspective holds that the world political economy ought to resemble—in Helleiner's terms—a neomedieval structure, where self-regulating local communities run their own economies, regulated by decentralized institutional arrangements. Under such arrangements, a global civil society would control the worst global environmental problems. The intellectual lineage of such ideas dates back at least to Adam Smith's descriptions of a decentralized, de-industrial world, and to economic and social thinkers such as Leopold Kohr, E.F. Schumacher (who was heavily influenced by Kohr), and Ivan Illich. According to Helleiner, many contemporary ecological and development economists have noted their intellectual debt to Kohr in particular and to ideas that flowed from his work.

From these observations it should be clear that the story of which economic ideas were selected and how they influenced the evolution of international environmental norms is not simply a matter of which ideas are "better" or which merely existed within a particular community of economists. The reasons why, after thirty years, ideas associated with liberal environmentalism gained a strong foothold must be analyzed historically to see the social and political processes through which that selection occurred. Clearly, the perspective of this book is that economic ideas mattered. This chapter shows which ones mattered and seeks to understand why.

The chapter demonstrates the influence of economic ideas, but not simply as embodied through an "epistemic community" of economists and like-minded policymakers. I have already shown that portraying economists as an epistemic community in the context of global environmental governance would undermine the explanatory power of the hypothesis, which relies on the privileged position of one particular group of knowledge-based experts to explain outcomes. More importantly, the empirical story of the role of economists simply does not fit an epistemic communities explanation: an identifiable transnational group of economists did not emerge to put forward or promote the sustainable development concept or the ideas behind liberal environmentalism. Most of the work by economists occurred in the absence of an organized transnational research program, at least until the move toward liberal environmentalism was well underway. When ideas did flow transnationally, they did so through governmental institutions that, while dominated by the economics profession, could not properly be considered an independent community of experts who had regularized contact, built consensus around an issue, and then promoted that issue independently based on a set of values. If anything, the advent of liberal environmentalism empowered economists in environmental policymaking, but could not be said to have arisen through their efforts.

Understanding what happened in the mid-1980s to move international norms toward liberal environmentalism thus requires looking far beyond the expert communities that studied such ideas. Instead it requires a wider look at ideational influences and how they interacted with powerful international norms.

The socio-evolutionary explanation for normative development put forward below takes this approach. This explanation shows how the ideas associated with liberal environmentalism interacted with an existing international social structure of governance to institutionalize the norm-complex of liberal environmentalism described in chapter 3. I apply the approach by tracing through the economic ideas that evolved into "sustainable development" thinking and eventually became institutionalized as liberal environmentalism. Thus, descriptively, the chapter shows the importance of economic ideas as the ideational basis for liberal environmentalism. In terms of explanation, the socio-evolutionary approach shows why those ideas prevailed over alternatives. It also leads to a better understanding of why international environmental governance evolved as it did since 1972.

SOCIO-EVOLUTION AND GOVERNANCE

Building a theory to map the interactions of ideas and the structures they encounter (however defined) has proven elusive. Whereas a number of scholars now recognize the utility of examining the role of ideas to explain the content of international governance, they have expressed frustration on two counts: either the literature does not adequately explain why ideas possess a causal power of their own or, if it does, then it fails to explain why some ideas are selected over others (Yee 1996; Blyth 1997; Woods 1995). What bedevils theories of ideas, and especially of how ideas become institutionalized as norms, is how to reconcile the interaction of the sources of ideas with structural explanations, where ideas and institutions are residuals of powerful actors pursuing their interests. Solutions inevitably address how ideas interact with their "environment," usually conceived of as institutions of one sort or another or as constellations of interests.

For example, Garrett and Weingast (1993:176), following Thomas Schelling's insights, make the rationalist argument that ideas create "focal points" around which behaviors of actors converge. The environment in which such ideas must operate is a constellation of actors with given interests. Ideas select from one of multiple cooperative equilibria available to create stable institutions. However, this formulation cannot explain the actual content of governance—or the focal point chosen—because any number of ideas would seem to do. In addition, this explanation remains extremely inefficient, requiring constant reevaluations of changing preferences over time.

For example, liberal or unit-based international relations theory would look to domestic theories of preference formation and change as a necessary step in such an explanation. Institutions then result from a configuration of preferences of states (Moravcsik 1997:537). By ignoring the constitutive aspect of international institutions, however, liberalism misses the potential starting point of using social structure to define why some preferences are viewed as more "legitimate" than others.[4] An explanation that endogenizes interests to some degree would be more efficient.

What is required, then, is to move beyond a rationalist approach that views ideas as simply intervening variables between given interests and behavior. The constructivist literature in international relations presents one alternative. It recognizes that interests themselves are derived, at least in part, from an existing normative or social structure in which actors participate (Dessler 1989; Wendt, 1992, 1994; Wendt and Duvall 1989; Kratochwil 1989; Busumtwi-Sam and Bernstein 1997). Constructivism focuses especially on how actors' interests derive from their "identities" (as sovereign states, or members of communities, and so on). In so doing, it emphasizes the constitutive dimension of norms, wherein norms do not merely regulate behavior, but define social identities and practices. Such practices are prior to individual action in that they define the range of meaningful if not conceivable behavior. However, constructivist writing to date has generally lacked a clear theory of how new ideas interact with the existing social structure. Such a theory would identify the process through which ideas become institutionalized as norms—in other words, how norms and social structure evolve.

A few international relations scholars, influenced by the "new institutionalism" literature,[5] have begun to address this interaction between ideas and social or institutional structures. For example, Steve Weber (1994) has argued that the creation and function of the European Bank for Reconstruction and Development can be better understood in reference to its "institutional" environment rather than its "technical" environment. The Bank, he argues, developed from a shared set of ideas and purposes of states in Western Europe (around the state, democracy, and market economics) and pushed "the ideas and consensus around them substantially further as part of an effort to extend east an ideational and institutional foundation for multilateral cooperation" (1994:2). He draws on organizational theory and the new institutionalism to suggest that the norms and functions of the Bank were less dictated by efficiency and means-end rationality (the technical environment) and more by their "social fitness" with existing institutions and political economic norms existing in Europe:

> Organizations in an institutional environment are judged by the appropriateness of their form; they compete for social fitness . . . and they are rewarded for establishing legitimate authority structures and procedures more than for the quantity and quality of what they produce. Ends and means are not treated separately, so that proper procedures and a "rationale"—an account that makes what the organization does understandable and acceptable within its social context—are the basis of legitimacy (1994:7).

Similarly, Jeff Checkel makes the case that changes in Soviet foreign policy under Gorbachev were made possible by a confluence of factors that included a reformist general secretary, a group of entrepreneurial purveyors of new concepts and ideologies, and "institutional and political settings that at different times either constrained or enhanced [entrepreneurs'] ability to influence policy" (1993:273). What these two studies have in common is that neither explains outcomes as merely dependent on the introduction of new ideas interacting with a set of existing interests in a rational manner. Rather, the normative context—or social fitness—of new ideas plays a key role.

Here I take the basic insights from studies such as these and from constructivism, to push the idea of "social fitness" further. What is required, I argue, is a clearer specification of social structure and a method to think about how ideas interact with that social structure—in other words, the selection process—to see the means through which some ideas get institutionalized as norms while others do not.

In contrast to the rationalist approach to ideas, I propose a *socio-evolutionary* approach that focuses on how ideas interact with existing norms, which are based in intersubjective understandings among actors. The explanation thus begins not with actors or state interests (as do liberal and rationalist explanations), but with systemic social structure. The explanation argues that three factors determine the selection of new norms: *the perceived legitimacy of the source of new ideas; fitness with extant international social structure; and fitness with key actors' identities at various levels of social structure.* By identities I mean both their status as agents as constituted by international social structure and their socioeconomic identities generated domestically which they project in their international affairs, such as their view of legitimate political and economic order as reflected in domestic institutions. I discuss each of these three factors in turn.

The approach is "evolutionary" because it identifies a selection process based on the interaction of ideas with their environment that has some parallels to processes of natural selection.[6] Ann Florini has thus described the acceptance of new norms as follows:

> Given two contested norms, one may be more prominent in the norm pool, more compatible with other prevailing norms, and/or better suited to the existing environmental conditions than the other. If so, that one will become more frequent in the population relative to the other (1996:369).

This analogy should not be carried too far, however, since unlike natural selection, the evolution of norms is a conscious activity that, while manifested in practices, exists in the minds of actors who engage in those practices. Human beings engage in purposeful action and self-reflexive thought. To quote Anthony Giddens, "every social actor knows a great deal about the conditions of reproduction of the society of which he or she is a member . . . institutions do not just work 'behind the backs' of the social actors who produce and reproduce them" (1979:5). Thus the process I identify does not simply result from survival or success of some traits over others that occurs independently of actors' understanding of them.

Hence a key component in normative evolution is the ability of purposeful actors to gain legitimacy for the ideas they promote through legitimating institutions, that is, institutions viewed as appropriate or legitimately engaged in the task at hand (the first causal factor identified). Therefore, I do not contradict the claim of the epistemic communities literature that legitimacy of knowledge claims can be an important causal factor. However, by focusing on the institutional basis of legitimacy, my explanation highlights that legitimating institutions do not necessarily reflect expert knowledge. The focus on processes of legitimation also avoids dependence on random variation or mutation as the source of change in my theory.[7] Thus, the mechanism of change has no direct parallel in evolutionary theory. It is the mixing of purposeful action and social structure that requires mapping.[8] So while I borrow the concept of fitness and modify it to mean *social* fitness, new ideas in the minds of actors and the legitimacy they can gain for such ideas are the drivers of change. They are not random ideas, but attempts, in essence, at the social construction of reality.[9]

Second, the "socio" part of the approach is the recognition that norms interact with a "social" structure of existing institutionalized norms that have already become legitimated; they exist intersubjectively as social facts and obligate because of agreement of members of the relevant community.[10] The process of institutionalization involves the nesting of norms, that is, when they become linked with other norms that express similar values, interests or goals (Busumtwi-Sam and Bernstein 1997). The key to the selection process is *social* fitness with already institutionalized norms, which constitute the "environment" new norms encounter.

The basic contours of social structure can be described in order to identify general patterns of institutionalized norms in the international system. These institutionalized norms, in essence, make possible meaningful action (Bhaskar 1979; Giddens 1979). Structuration theory and constructivist international relations scholarship emphasizes that social structure results from the interaction of practices of the units it comprises. Once institutionalized, however, those practices exhibit structural qualities in that they legitimate practices consistent with their logic and marginalize all other practices that pose a challenge (Doty 1997). Social structure evolves in response to new norms while it also reproduces practices (of states and other relevant actors) that new norms attempt to alter. In this way, social structure constrains and enables the entrance of new ideas. Norms that do result from this interaction must not only come from "good ideas," but must draw on rules and resources enabled by existing social structure.

The content of international social structure can be conceived of as *institutionalized* norms ordered in a hierarchy of prioritized values that coordinate and define international interactions. This formulation follows from a number of scholars who make reference to levels of norms.[11] Although authors conceive of levels in various ways (for example, by functional domain, general versus specific institutions, and so on), viewing them as arranged in a hierarchy is useful in order to depict social structure as an authority relationship. As a source of authority, social structure is also a structure of governance. In other words, political authority rests on the institutionalized norms that constitute social structure and thus define which political institutions and practices are viewed as appropriate. Social structure constitutes a hierarchy of authority in that norms at deeper, or more fundamental levels condition or make possible (constrain and enable) other less fundamental institutions (Wendt and Duvall 1989:64).

This constitutive aspect of international norms and rules has important implications for the study of international institutions. It means they do not simply rise and fall in response to the will or interaction of preexisting states (Ashley 1984). Rather states and institutions are both constituted by a normative structure that, in a sense, goes all the way down.[12] Moreover, the institutions that rationalists tend to focus their analysis upon themselves are enabled by norms that constitute states as actors who can negotiate treaties or cooperate to create them in specific issue areas (Dessler 1989). Hasenclever et al. (1997:158–159) sum up this argument based on early constructivist scholarship:

> The norms and rules that make up the institution of sovereignty define intersubjectively who can lay claim to the status of an actor in international politics and what rights and duties each actor bears in principle. Without such norms and rules it would make no sense to speak of either illegal intervention or legitimate self-defense. . . . Similarly, the norms and rules specifying the legal principle of *pacta sunt servanda* [treaties must be served] define what counts as a treaty and determine the conditions under which a treaty has to be regarded as binding the future volition of the participants. . . . And this is more than a semantic statement: in the absence of such rules, practices such as treaty-oriented negotiating or treaty-stabilized cooperation would not exist as well.

Following from these arguments, international social structure can be thought of as consisting very generally of three basic levels, with more fundamental norms at deeper levels than less fundamental norms. At *level one* are the fundamental norms of identity—the deep structure of the system—which identify who or what the primary actors are. Constructivists generally agree that norms that define "who counts as a constitutive unit of the international system" are the deepest (Ruggie 1998:20; Kowert and Legro 1996:467–468). In the contemporary period, constitutive norms have institutionalized sovereignty as the organizing principle of the international system and identified states as the sole repositories of sovereign authority.

At *level two* are norms that specify the minimum conditions for the co-existence of sovereign states. Level two norms specify the basic obligations owed to the society of states as a necessary and reciprocal requirement of membership. Norms at this level impart a particular social meaning to sovereignty. In other words, these norms perform the primary function of defining and stabilizing property rights by specifying how actors who claim the status "sovereign" are differentiated from other similarly constituted actors and how they may or should act.[13] For example, these norms specify the sovereign equality of states, rules of nonintervention and nonaggression, property rights of states and jurisdiction, and diplomatic immunities. They also specify derogations from absolute sovereignty required for coexistence. In the contemporary state system, the key norm at level two is the notion of general and diffuse reciprocity. That is, all states are to be treated essentially alike, and enjoy the same rights and responsibilities. The norms of reciprocity and multilateralism, embedded in many less fundamental institutions, thus fall at this level.

Finally, *level three* norms constitute and regulate social relations differentiated functionally—by activity—as actors collaborate and coordinate activities to manage interdependence and solve collective action problems in specific issue areas, such as international commerce or environmental management. Such institutions by definition limit the *autonomy* of state action, but rest on state *authority* at deeper levels of social structure.

The exact content of each level is less important than the argument that social structure as a whole functions to arrange rules and institutions in terms of who/what is empowered to act, how they may act, and the range of actions they may legitimately take in pursuit of interests and objectives.

New ideas in one issue area are more acceptable if they "fit" with other norms at level three that govern related issue areas where functions may be linked or interdependent. In organizational theory, Zucker (1991:105) has

referred to this process as the "contagion of legitimacy" where norms become linked in networks of related functions so that changes in one neces-
sitate changes in others; thus change is resisted. For example, in this case,
new ideas about environmental practices are more acceptable the more
compatible the fit with norms that govern interactions in related issues,
such as trade norms embodied in the GATT/WTO. Contestation results
when they do not. Indeed, much of the contestation over global environmental norms occurs at this level. Although a social structural explanation
cannot determine for certain the outcome of that contestation, agreement
on norms that conflict with other norms governing related issues at level
three is likely to be more difficult and less coherent, and agreements that
do result are often shaped by existing norms.

However, fitness also requires compatibility with norms at deeper levels.
The formulation of international social structure here means that level three
norms must be compatible with norms at levels one and two for social fitness to be easily achieved. For example, the evidence will show that evolving
norms of sovereignty and multilateralism favored some forms of global environmentalism over others. Social structure affects barriers to entry and
exit, political capabilities and external legitimacy claims. It favors some actors and identities over others, and thus shapes and shoves self-interest,
which becomes defined in ways consistent with favored identities, and the
ways to get what one wants within the social structural setting.[14] Seen in this
light, the evolution of environmental norms demonstrated in earlier chapters can be seen in the context of their fitness or competition with norms at
various levels of social structure. The important point here is simply that
new norms are likely to face much greater contestation if they are incompatible with deeper norms, such as those that specify core actor identities.

For the purposes of my socio-evolutionary explanation, I am not proposing a full-fledged theory of social structure. Rather, with this general
notion of social structure in mind, I argue that its basic contours can be
identified at any given time to identify the context in which new ideas must
interact. Thus, while recognizing that systemic social structure itself is constantly evolving owing to the interaction of practices of actors (mainly sovereign states[15], but other actors as well), that structure can be held constant
for the purposes of analysis and identified as the "environment" that new
norms face. In other words, social structure, at any given time, has a determinative content (norms and institutions), which can be gleaned through
careful historical analysis and informed interpretation.

Because norms constantly face contestation, social structure is not completely determining. Actors make choices, constituted and regulated by

rules afforded them by social structure, which may delegitimate as well as legitimate norms and institutions. In recognition of this co-constitution of agents and structures, the final explanatory factor identified—compatibility or incompatibility of new norms with state identities—is brought in not as an entirely independent variable from social structure, but to highlight that agents' identities and interests simultaneously constitute and are constituted by social structure. For example, state interests derive in part from their "identities" as sovereign states, or members of communities such as "developing countries" or the "G-7," which then contribute to how these actors understand their interests. In addition, some fit with domestic political and socioeconomic organization among relevant states (their internally generated identities), and the way they project those values internationally, is also required since international social structure in the sovereign state system privileges states as actors (Ruggie 1983). A focus on social structure does not negate the role of domestic factors. However, even when interests are defined by reference to internal factors, they are also mutually constituted in a context of interaction governed by rules. The analysis can therefore remain primarily at the social structural level since I make no claims about the ultimate source of internal identities for the purposes of this explanation.

Power therefore plays a role in the reproduction and transformation of social structure, although not as traditionally measured by material capabilities alone. Rather, power is exercised in the context of an existing normative structure that reflects practices of actors constrained and enabled by that structure: agents operate within systems of rules that empower them and endow material resources with meanings and uses (Wendt 1994). In other words, power matters, but norms contextualize power relationships; they are the media through which actors decide on the appropriateness of practices and engage in contests over those practices. In this vein, John Ruggie (1983) argues that one must examine how power (capability) and social purposes become fused to project political authority, thus contributing to the formation and maintenance of international governance structures. In a normative structure that privileges sovereign states, the introduction of new ideas still depends in part on congruence with their social purposes. If identities of states change for domestic reasons, those changes can produce contestation within international social structure. Shifts in social purposes in major states may therefore signal changes in social structure more broadly.

It follows that social structure is *emergent*: agents act reflectively and "can produce instabilities that generate the propensity for self-organization" into new self-reproducing structures.[16] This process is especially likely when new problems or circumstances arise. Any social structural explanation is therefore limited in its ability to predict in the mechanical sense, that is, to specify generalizable antecedent conditions for particular outcomes because that structure itself evolves historically. However, by providing *reasons for action*—collective intentions and aspirations, legitimate behaviors, and so on—the general contours of social structure can be identified and held constant as described above to establish the available set of rules and resources on which actors can draw to construct new meanings (Ruggie 1998:16–22; Yee 1996). As the content of social structure changes, so too do the conditions for new norms to gain legitimacy. In this way, ideational change is contingent on social structural factors, and may lead to unintended consequences.[17]

Applying the Explanation

The remainder of the chapter shows that a socio-evolutionary explanation better accounts for the shift in environmental norms than alternatives already examined. Applying a socio-evolutionary explanation involves tracing through attempts to introduce ideas that linked environment and development, showing that actors promoted those ideas through legitimating institutions, and assessing how extant social structure selected some ideas over others and shaped the norms that were institutionalized toward liberal environmentalism.

Empirically, I show that economic ideas overshadowed scientific ideas and ecological thought in producing normative compromises at key junctures in the evolution of the environmental norm-complex over the last thirty years. Following the pattern delineated above, economic ideas that became influential largely did so within key institutions that, owing to their legitimacy, disseminated those ideas to member governments and other international institutions. Specifically, policy entrepreneurs working through the OECD and Brundtland Commission played the most significant role in generating economic ideas that brought environmental concerns into the mainstream of international governance. The OECD also identified the compromise—the fit with international social structure—that would eventually form the basis of agreement at UNCED. Policy entrepreneurs succeeded best at moving a concern for the environment into the mainstream of international politics when they nested environmental norms into the broader international social structure, even as that structure evolved to reflect a new consensus on a liberal economic order.

Without these ideas and their nesting within key institutions, environmental governance was unlikely to have developed as it did and agreement on norms would have been more difficult and less coherent. Legitimacy of ideas mattered, but not simply in the way an epistemic communities hypothesis would suggest. In this way, the separation of economic and scientific ideas highlights the way a socio-evolutionary explanation avoids privileging particular ideas in advance, as the epistemic communities literature does with the argument that a single community is granted legitimacy based on its claim to authoritative and policy-relevant knowledge in a given issue area. In this case, that would leave the compromise of liberal environmentalism unexplained. Although an epistemic community is potentially a source of new ideas in both explanations, the difference comes in the starting points, and thus the main causal mechanism at work, in that a socio-evolutionary approach begins with social structure, through which ideas are selected, whatever their origin.

I have already presented much of the empirical detail for this explanation in chapters 2 and 3. To avoid repetition, below I elaborate on key events in the evolution of environmental norms only to provide evidence for the socio-evolutionary explanation, and refer back to earlier empirical details as appropriate.

LIMITED SUCCESS: ECONOMIC IDEAS, THE NORTH-SOUTH DIVIDE, AND ECODEVELOPMENT

The story of attempts to bring environmentalism into the mainstream of the international agenda demonstrates the repeated influence of economic ideas, but with varying speed and success. However, many of these economic ideas were not fully accepted until the advent of sustainable development thinking. When ideas succeeded, entrepreneurial leaders such as Maurice Strong, Mustafa Tolba, Jim MacNeill, and Gro Harlem Brundtland were necessary to pull those ideas together in legitimating institutions. When they did not fit with dominant social structures or make headway within legitimating institutions, those ideas failed to become institutionalized.

Beginning with Stockholm, recall that it was development and environment economists, not natural scientists, who first introduced the environment/development linkage that Stockholm Secretary-General Maurice Strong recognized as a requirement to get many developing countries to participate. These ideas responded to developing states' fears about an emphasis on lifeboat ethics or no-growth philosophy implied in studies such as *Limits to Growth* (de Almeida 1972; *Founex Report* 1972:12–13, 27). In particular, developing countries worried that trade barriers would be erected under the guise of environmental protection.

As mentioned in earlier chapters, Strong convened two key meetings in 1971 of small groups of "experts" to respond to these concerns. It was the second meeting, primarily of developing country economists in New York, that he called "the single most influential meeting in terms of my development of the agenda" (the first meeting at MIT was described in chapter 4).[18] According to Strong, the New York meeting specifically aimed to bring development onto the Stockholm agenda. As former head of the Canadian International Development Agency, he had already decided that environment and development needed to be linked, but asked economist Barbara Ward to convene a group of "developing country economists and thinkers to really address these issues." As he put it, "it didn't take a genius to figure out that through the development process that we affect the environment, and only through improved management of the development process that we can actually address realistically environmental issues." The New York meeting of economists started the process of giving that environment/development linkage intellectual content and legitimacy, especially in the developing world:

> I was world champion right from the beginning of the whole need to integrate environment and development. That was my whole thesis for coming into it, my first speech to the preparatory committee made that clear. Now, mind you, that was a simple conceptual approach. I needed to put flesh on the bones and I needed a lot more professional guidance and professional help, and I of course went out to seek that. I'm not saying that I invented that, I mean that was such a logical thing. But I did in fact re-orient the entire Stockholm agenda [from the focus on pollution only, under Swiss scientist Jean Moussard].

Thus, it was the New York meeting that really allowed Strong to formulate ideas that would get developing countries interested in Stockholm, and could be considered the beginning of the evolution of ideas toward what eventually became sustainable development thinking at the level of international discourse.

Social structural factors, however, set the context for such efforts on the part of developing countries. At that time, developing countries' group self-identities, as reinforced through organizations such as the Group of 77 (G-77) and the UN Conference on Trade and Development (UNCTAD), reflected a view of the liberal economic order as structurally exploitative of them. These identities reflected in part their relative positions in the world economy, but also rested on shared historical experiences of colonialism and ideological commitments to economic theories such as dependency that gave a coherent, intersubjective, understanding of global order (Krasner 1985:53–60). Hence norms that underlay developing country positions stemmed from these identities and were reflected in a more general program of development goals. Developing countries, or experts from those countries, formulated and expressed these goals in relation to the environment in two more formal meetings in the lead up to Stockholm: the November 1971 Second Ministerial Meeting of the G-77 in Lima, Peru and the meeting of development experts in Founex, Switzerland. Founex, as mentioned in chapter 2, made a much more direct impact on the Stockholm preparations and, according to Strong, "it arose directly out of the [smaller] meeting . . . in New York," which also provided Founex's intellectual foundation.

Founex's significance is worth reviewing. First, the report it produced demonstrated that developing countries were concerned about environmental problems, but were deeply suspicious of how the international community would deal with such problems if treated in isolation from development. Second, the report differentiated the environmental concerns of developing countries from developed countries. Finally, it presented environmental concerns in the context of a set of international norms consistent with developing country concerns, in particular those expressed in the Strategy for the Second Development Decade. For example, it emphasized the sovereign control of developing countries over their economic development and their own resources (*Founex Report* 1972:11, 22). In brief, Founex juxtaposed development and environment, showed skepticism for Northern concerns over global environmental problems when local environment and development problems related to poverty seemed far more important, and demonstrated a general antagonism toward a liberal economic order that appeared to unfairly disadvantage developing countries.

This combination of ideas meant that the economic ideas contained in Founex were unlikely to forge a global consensus acceptable to the North. Nonetheless, Founex forcefully demonstrated the need to reconceptualize the new international environmental agenda if a truly global consensus on action were to be achieved. Many of the economic ideas contained in

Founex persisted in various developing countries and cooperative fora right up until UNCED. For example, economic concerns expressed in the 1991 South Centre report show remarkable similarities to those in Founex. Many developing countries at UNCED, however, moved from those positions toward the consensus around norms in the Rio Declaration. Had new concepts not developed to forge politically acceptable linkages between environment and development, environmental governance might not have progressed beyond this North versus South, environment versus development stalemate.

The concept of ecodevelopment promoted at UNEP under Strong's leadership in the 1970s constituted the first real attempt to create a unifying set of ideas around environment and development. It in many ways came closest to the kind of synthesis that Haas's scientific ecologists might have desired. However, this attempted "marriage" of ecology and economy did not come from the science of ecology *per se*. Notably, as I pointed out earlier, Strong credited Ignacy Sachs, a French development economist, with giving the concept intellectual content, although ecological ideas clearly had an influence.[19] However, the concept did not translate into great success in shaping governing norms (Caldwell 1990:202–204; Adams 1990:51–56).

Earlier I detailed the practical problems with ecodevelopment. However, another problem was that the concept itself was redefined from UNEP's attempt to fit ecodevelopment into a program that broadly promoted economic growth, into something more radical (Moffatt 1996:10). Robert Riddell's (1981) influential reformulation of ecodevelopment in his book of the same title—the most commonly cited on the topic—shows a much deeper suspicion of economic growth. It views growth through global economic expansion and industrialization as almost inevitably perverse in that it benefits already wealthy nations and small elites in developing countries. Furthermore, it perpetuates poverty, unemployment, overuse of resources, economic dependence, and a host of other social, economic, and environmental problems that hurt the majority of people in developing countries. The subtitle of Riddell's book—"Economics, Ecology and Development: An Alternative to Growth Imperative Models"—reveals the general orientation of his thought, which, although not hostile to growth *per se*, emphasizes "progress more than growth" (1981:149). Progress can be achieved, he argues, through administrative and social reform that includes population control, increased self-reliance at the individual and community levels, improved nutritional quality of food intake, and various other small-scale and local development initiatives. These programs should aim to fulfil local needs, and the achievement of social and economic goals in the context of resource conservation and environmental protection.[20] Growth through economic expansion, large-scale capital investment, and liberalized trade is thus inimical to ecodevelopment as formulated by Riddell. This radicalization of ecodevelopment did not sell well in developing countries, leading politicians and industrialists did not support the concept, and the proposals appeared to many as idealistic and impractical (Moffatt 1996:11; Adams 1990:55–56).

Strong attributes this lack of success to the more general negative reaction in developing countries to ecodevelopment's apparent ecological focus. As Strong put it:

> I suppose there's still a sense in which many people regarded the word "eco" more in its ecological than in its economic sense and the word sustainable appealed more to the sort of outright development constituency. The word "eco"—we never succeeded in getting it across as a synthesis of the two ecos. Rather it seemed to come out more on the ecological side. . . . I think in effect sustainable development sold better in the development side of the constituency (author's interview).

Neither the scientific and development communities around UNEP, nor the economists and other development experts in international development agencies, could produce the necessary North-South consensus or alliances among key state or institutional actors to create a focal point for environmental governance around ecodevelopment. Similarly, the World Conservation Strategy had only limited success in shaping overall governance structures, further demonstrating the limitations that policy-oriented scientists and environmentalists worked under when trying to sell environmental concerns as compatible with development.[21]

Although ecodevelopment failed in large part because of its incompatibility with economic norms at level three of social structure, it also posed a challenge to norms at levels one and two. Its failure exemplifies the difficulty new ideas face in becoming institutionalized when they challenge deep norms of social structure. Along with related strands of "green" thought, ecodevelopment stressed local control of economic activity and decision-making that could diminish the administrative control of the state. Recall that developing countries at Stockholm successfully institutionalized sovereign control over natural resources drawing on deeply institutionalized meanings of sovereignty in the postwar period. Given the limited lasting influence of other NIEO proposals, the legitimation of this norm provides evidence that it stood apart because it fit better with extant international social structure at the time. Since World War II, level two of social structure has emphasized "negative sovereignty"—that is, sovereign equality and self-determination as reflected in state practice has been defined as freedom from external domination or anti-colonialism (Jackson 1990). In this context, ecodevelopment's relative failure is understandable, as is the dashed hope of some activist scientists or environmentalists for a more holistic or global approach at Stockholm. Deeper levels of social structure militated against responses that appeared to undermine sovereign authority or centralized adminstrative control of the state.

The failure of ecodevelopment is an important part of the normative story since the failure of governing norms also gives insight into why others succeeded. As Moffatt (1996:11) colorfully puts it:

> One of the positive results stemming from the theory of ecodevelopment . . . was that sustainable development became considered as an alternative way of organizing socioeconomic development in a way which would, as far as possible, result in less harmful environmental practices. Like the parable of the sower, many seeds of these ideas fell on barren ground; some were cultivated as ideological blue or red blooms, and fortunately, one or two seeds were able to flourish . . . [as] in the Brundtland Report.

The theoretical issue is why ideas associated with the Brundtland Report, and norms that followed from it, succeeded where others had failed.

SUSTAINABLE DEVELOPMENT AND THE OECD

The pivotal change in framing the problem of environment and development came from the OECD Environment Directorate. In this case, the OECD acted not simply as a club of rich countries, but as a research organization and policy entrepreneur. Its development of the Polluter Pays Principle (PPP) in the early 1970s created an intellectual basis on which to build future policies (OECD 1975). As mentioned, the spirit of the principle implies the use of market-friendly instruments such as pollution charges and tradeable permits that have been the trend in implementing the PPP in the 1980s and 1990s. Its main purpose is to support economic growth by achieving environmental protection with minimal distortion of markets. The notion of 'getting prices right' underlies the principle.

Given the growth-oriented goals of the OECD and other lead institutions of intergovernmental environmental governance (for example, the EC/EU and later the World Bank) the emphasis on growth- and market-oriented environmentalism within those institutions is not surprising. Hence, a focus on the normative environment within the OECD (the organization itself as well as member states) makes the policies outlined in chapters 2 and 3 more understandable. For example, Article 1 of the OECD Convention (signed December 16, 1960 in Paris) commits the organization to policies that aim "to achieve the highest sustainable economic growth . . . and sound economic expansion" of members and non-members and the expansion of multilateral, non-discriminatory trade (OECD 1973a:48). Following these goals, its stated environmental position recognizes "that governmental interest in maintaining or promoting an acceptable human environment must now be developed in the framework of policies for economic growth" (1973a:15).

Indeed, the OECD pioneered many aspects of economic thinking about the environment. These efforts came mostly from its environment committee established in 1970. There, a core subcommittee of economic experts introduced many of the ideas that the OECD council would later adopt and push in member states and at international gatherings (OECD 1973b:23). For example, the subcommittee developed the "Guiding Principles Concerning International Economic Aspects of Environmental Policies," of which the PPP is a cornerstone. These principles fit within the committee's primary mandate to "investigate the problems of preserving or improving man's [sic] environment with the particular reference to their economic and trade implications." Its guiding philosophy is that only an "expanding economy can provide the resources to meet the higher expectations of man [sic] in his quest for a better quality of life" (OECD 1973b:7–8). Complementing this pro-growth orientation, committee-sponsored work uses cost-benefit analysis as its primary method of evaluating alternative environmental proposals. Its research has concentrated on problems such as how to implement the PPP while maintaining fair competition in different jurisdictions. Current work in the Environment Directorate continues along these lines, and, since Brundtland, especially focuses on the use of economic instruments.[22] However, these ideas did not just drop into international discourse or automatically become entrenched in environmental norms. Rather, individuals within OECD had to mobilize these ideas in order to pull the weight and legitimacy of the organization behind the dissemination of those ideas to powerful actors.

Similarly, the ideas germinated within the OECD for varying lengths of time before member states began to shift their policies toward this view of the environment, and thus they cannot simply be seen as deriving from domestic politics of powerful states. The ideas promoted transnationally clearly interacted with policy experiences and socioeconomic changes within member states, but the OECD as a policy organization played a leading role in defining and promoting particular policy responses ahead of the policy transition in member states (Hajer 1995; Weale 1992).

The signal event that brought these ideas into the mainstream of public policy came in 1984 when then Environment Director Jim MacNeill organized the "Environment and Economics" conference. MacNeill felt that part of the reason the decade since Stockholm produced such a poor environmental record was that it focused on "end-of-the-pipe" solutions that were a cost-burden to industry. The Environment and Economics conference meant to provide an economically rational response to such problems. In MacNeill's view, "We laid the intellectual foundations for what later became known as sustainable development in [the] OECD between '80 and '84," and the Economics and Environment conference articulated that foundation.[23] It greatly influenced member governments and business leaders, who were well represented there, in the direction of the OECD Environment Directorate's vision of environmental governance. The conference included high-level participants from OECD member governments, many at the ministerial level (for example, U.S. EPA Administrator William Ruckelshaus chaired one of the sessions). Also present were members of the European Commission, leading academics from the environmental economics community (David Pearce, for example), interested governmental and nongovernmental organizations, trade unions, industry leaders, and prominent individuals including Maurice Strong. The conference was chaired by Pieter Winsemius, minister of housing, physical planning, and the environment of the Netherlands. Recall Winsemius later co-authored, with MacNeill and Taizo Yakushiji of Japan, a major report on this topic for the Trilateral Commission (MacNeill, Winsemius and Yakushiji 1991).

Not only did the conference disseminate its views to powerful leaders in government and industry, it also constituted a learning process for OECD economists. As MacNeill put it, "[The conference] was with the active participation of the economic establishment in the OECD, and you know the OECD is to classical economics what St. Peters is to Christianity. I mean it's the keeper of the keys. And so we had the economic establishment involved and that was essential" for a wider policy impact (author's interview).

In a sense, this process of how new ideas found a fitness within the constraints and opportunities of the normative environment of the OECD is a microcosm of the larger process of normative evolution. Thus, the socioevolutionary approach taken here clearly has similarities to arguments made in the "new institutionalism" literature to the effect that the "ideas and intellectual outlooks of specialists are filtered through the institutional settings in which they operate. Depending upon the details of a particular organization's history and sense of mission, these settings can either hinder or promote particular ideas or outlooks" (Checkel 1993:277–278; March and Olsen 1984:739). New ideas about environmental policy put forward in the Environment and Economics conference were also shaped by the organizational goals and norms of the OECD and the economics profession that dominated its work. In addition, with the "economics establishment" within the OECD on board, legitimacy within the OECD as a whole was greatly enhanced.

The findings of the Economics and Environment conference emphasized the desirability of strengthening the role of economic instruments and the reciprocal positive linkages between environmental protection policies and economic growth. Studies presented found that expenditures on environmental protection had actually increased growth, spurred innovation, and increased jobs at the macro level, although losses might be experienced in specific industries. It also found that economic instruments were more efficient, more effective in the promotion of innovation, and more appropriate for environmental policies that had shifted toward prevention. This latter finding was key, as it suggested that not just any form of environmental protection could solve the environment/economy dilemma, but policies that geared environmental protection towards compatibility with economic growth and the operation of the market would (OECD 1985; Jim MacNeill, author's interview).

These findings, MacNeill said, turned conventional wisdom on its head that said the environment and economy were enemies, and that the best that could be achieved was a balance between the two. He summed up the influence of the conference this way:

We came out of that conference with [the] OECD saying for the first time . . . that the environment and the economy can be made mutually reinforcing. That was a breakthrough conclusion for that organization. It changed the way a lot of people began to think about the environment and the economy. And it was that conclusion . . . that I took with me into the Brundtland Commission in late '84 when we got started.

And that formed one of the cornerstones, and a very large cornerstone, for the Brundtland Commission's report and its conclusions with respect to sustainable development (author's interview).

In this way, the conference played a major role in shifting the way governments, business, and the economic establishment at the OECD thought about environmental issues and the best ways to address them. In particular, the conference cemented the view, at least among key elites in the North, that economic growth and environmental protection could be compatible.

MacNeill's later role as secretary-general of the Brundtland Commission—Gro Harlem Brundtland picked him based on her contact with MacNeill while she served as Norway's representative to the OECD Environment Committee—ensured these ideas would gain further legitimacy. In addition, the Brundtland Commission could shape those ideas to find synergies with other norms then dominant in multilateral activities between North and South. Asked whether his work at the OECD had a big influence on the Brundtland Commission's work, MacNeill responded: "Oh, well I know that! I mean I brought it into Brundtland and I was primarily responsible for Brundtland, so sure, it had a big influence there" (author's interview). He also said the ideas influenced industry through groups such as Stephan Schmidheiny's Business Council on Sustainable Development.

MacNeill did more than just carry ideas, he wrote a detailed agenda for the Commission that would be distributed to all participants. However, the original version contained two agendas—the "standard" agenda, which fit with traditional approaches to conservation and environmental protection, and the "alternative" agenda, which—although it contained many other facets—fit generally with the OECD findings already listed. The other key innovation in the "alternative" agenda was to take a sectoral approach. That meant looking at environmental problems in the context of the economic sector as a whole in which they occur, rather than looking at a particular environmental problem (for example, natural resource issues such as deforestation or pollution issues such as acid rain) in isolation. A sectoral approach would, for example, examine the problem of climate change in the context of the energy and transportation sectors, and tropical forests and biodiversity would be approached from their common sources in agricultural, forestry, trade, and aid policies (MacNeill 1984:17–27).

In what was arguably the most significant decision of the Commission, it decided to distribute a version of the report to those making submissions that contained only the "alternative" agenda. That agenda then formed the intellectual basis for Brundtland's version of sustainable development.[24]

Asked why the Commission had not settled on some other terminology, such as IIASA's "sustainable biosphere," MacNeill said that it represented simply an updated version of the standard environmental protection agenda. As such, it would have led to proposals not much different than the World Conservation Strategy, which he felt just tacked development on to a resource management agenda. In contrast, "If you read [*Our Common Future*] you'll find that we begin with growth and the growth imperative. And we talk about the environmental consequences of that and we raise questions about the sustainability of growth. . . . So our point of departure is not the environment. It's the imperatives for growth" (author's interview). This selection process by MacNeill and the Commission helped to ensure that ideas embodied in the "alternative" agenda would dominate. In other words, it ensured that economic growth would from then on be at the core of global environmental governance.

The influence of the OECD on the Brundtland Commission admittedly worked somewhat in the way the epistemic community literature would predict. However, key actors who carried ideas were policy entrepreneurs, not primary researchers. More importantly, the ideas adopted rested less on consensual knowledge and more on institutional backing where legitimacy existed for the policy areas Brundtland would address thanks in large part to the groundwork laid by OECD. Policymakers took notice when the OECD, viewed as highly legitimate and appropriately engaged in economic matters, supported the environment. The UN backing of the Brundtland Commission provided added legitimacy for those ideas to gain a wider government and nongovernmental constituency in the North and South.[25] Organizations such as UNEP and IUCN had greater difficulty in gaining legitimacy for their ideas beyond their environmental constituencies.

At around the same time as the release of the Brundtland Commission report, many domestic environmental policies underwent changes that incorporated ideas consistent with the report's findings, even if actual implementation of such ideas has been uneven. For example, Weale and Hajer discuss these changes as a move toward "ecological modernization" and away from the previous conventional wisdom that the balance between the economy and the environment was a zero-sum game.[26] Most of these changes in domestic policies followed on the heels of the Brundtland Report, although strict causality would be difficult to determine as many of the changes occurred virtually simultaneously with the report's findings, and some occurred earlier. Nonetheless, Brundtland did provide a legitimating set of policy norms and responses to environmental problems that seemed to respond to the failures of policies in the 1970s. For example,

Weale notes that a large number of OECD countries began to recognize that old environmental problems continued, despite expensive regulatory measures, while new environmental concerns, particularly transnational issues, gained greater prominence (Weale 1992; Hahn and Stavins 1991). The responses advocated by the Brundtland Commission helped to reframe environmental policy discourse in a way that states and various interested publics found palatable as responses to such problems.

Although Weale (1992:31) points to a large number of domestic socioeconomic factors that contributed to these changes in domestic policies, he also notes that "the argument emerged, most notably in the Brundtland report, that environmental protection to a high level was a precondition of long-term economic development," and that domestic reforms drew from these new ideas of ecological modernization articulated therein:

> This body of ideas became appealing to many members of the policy elite in European countries and international organizations during the 1980s. . . . Part of its appeal, I conjecture, is that it has the potential to break the political stalemate between the clean air advocacy and the economic feasibility advocacy coalitions. Once it is recognized that pollution control can itself be a source of economic growth . . . then the balance of argument in terms of economic feasibility is tipped toward clean air rather than away from it (1992:79).

These new ideas thus "fit" relatively well with general economic goals of relevant states and also responded to a growing public concern. This discussion should not imply that the approach taken here better explains the actual direction of environmental policies in any given country than factors identified in alternative approaches to explaining domestic policy. Rather, it simply shows that the norms promoted by Brundtland found a fitness with the social identities and purposes of states around this period. As I will show presently, those social purposes themselves also began to undergo changes that coincided with changes in the international political economy as a whole.

THE "FIT" WITH SOCIAL STRUCTURE

The Brundtland Commission process coincided with a period of change in the international political economy and domestic economies in many countries. Internationally, remnants of New International Economic Order (NIEO) discourse still existed within the United Nations system, but any serious negotiations over an NIEO had already ground to a halt by the early 1980s. In this context, the Keynesian-style compromises described in chapter 2 reflected somewhat the broader normative shape of North-South dialogue in the mid-1980s. However, by 1987 when the Brundtland Commission released its report, those norms were already losing legitimacy. These changes in the international political economy in the late 1980s created an enabling environment that encouraged one particular path from Brundtland, reinforcing the legitimacy of the parts of the report most consistent with liberal environmentalism. The line of thought on which Brundtland based its core findings clearly legitimated this pathway, but the shift in economic norms created a particular social structural environment that selected how Brundtland would be used.

The key change that nearly coincided with the release of the Brundtland report occurred primarily at level three of international social structure, evidenced by changes in developing country identities away from consistency with NIEO goals toward, in many cases, identities more consistent with full participation in a liberal economic order. As described in chapter 3, these changes in the late 1980s, epitomized by the "Washington Consensus" to combat developing country debt, created a hospitable normative environment for WCED's ideas that formed the basis of liberal environmentalism. The general trend toward the retreat of the state from the economy, opening financial markets, promoting free trade, and acceptance of market forces as the main engine of economic growth gained wide acceptance in North and South alike, even in many formerly socialist economies.[27] I am not attempting to explain this shift in economic norms, only noting that given the broad experience of these changes at the global level, they would be difficult to derive from domestic preferences alone. Indeed, this change in social structure is an important story in itself that others have attempted to explain elsewhere.[28] For the full picture, one would need to combine the stories of the switch from Keynesianism to monetarism in economic policy in OECD countries, policy convergence in the European community, and the spread of the policy consensus through the developing world and the former communist bloc at the end of the Cold War.[29] The important point for the argument here is simply that such changes occurred, altering the social structural environment with which ideas about environmental action would interact. Most notably, the changes in social structure helped legitimate the framing of the environment problematique agreed to in Rio.

Interpretations of sustainable development during the lead-up to Rio emphasized elements consistent with these social structural changes. Recall that G-7 summit statements, later OECD environmental policy proposals, EC and EU environment programs, and a number of statements from international environmental gatherings demonstrated the impact of these ideas on future environmental policy research and programs, both domestically and internationally.[30] It should be noted that many of these research programs focused heavily on market mechanisms and fit with liberal economic norms that promoted growth. Thus, they emphasized the pathway from the Brundtland Commission report most consistent with the research generated at OECD.[31] MacNeill's own work in the period between Brundtland and Rio further entrenched this particular pathway to Rio with his influential report to the Trilateral Commission (MacNeill, Winsemius, and Yakushiji 1991).

In an even greater sign of just how far the normative shift reached, UNCTAD, the ideological center of demands for a NIEO in the 1970s, began in 1991 to work on greenhouse gas emission trading, and issued a major report on the subject in 1992 as a contribution to its work on the Earth Summit. As discussed in chapter 3, this reflected a broader normative shift within the institution and among its member governments, both toward a greater acceptance of liberal economic regimes and a greater attention to the environment. The 1992 "Cartagena Commitment," for example, explicitly reinforces the main Brundtland Commission finding on the compatibility of growth and environmental protection, and its analytic work supported the view that trade and sustainable development can be made mutually supportive (United Nations 1993:para. 39; Arda 1996:81). Its work on market-based solutions continues, including its organization, along with the Earth Council, of the Greenhouse Gas Emissions Trading Policy Forum to provide support to interested governments, corporations, and NGOs to launch a plurilateral GHG emissions market. It also coordinated research within the United Nations system on design and implementation of an effective Clean Development Mechanism for the climate change convention. More broadly, it continues to conduct focused research and support policies that promote positive linkages between liberalization, economic growth, and sustainable development.[32]

Economists often played a major role in influencing such programs, research, and policies through their analytic work, but became empowered by liberal environmentalism as much as vice-versa. As norms around liberal environmentalism gained legitimacy, policymakers and political leaders enlisted environmental and other economists to formulate specific policies that fit with the new-found legitimacy for growth-oriented environmental policies.

The U.S. case provides an excellent example. Project 88, mentioned briefly in chapter 3, came about not through the initiative of Robert Stavins, the Harvard economist who headed up the project, but through Senators Timothy Wirth (D-Colorado) and John Heinz (R-Pennsylvania). The two senators saw an opportunity to bring greater legitimacy to ideas that had been promoted already, with only limited success, by nongovernmental groups such as the Environmental Defense Fund. As Wirth put it:

> Senator Heinz and I thought that economics was pervading everything else during the Reagan era and a lot of other issues were being looked at through an economic lens and why should environmental issues be excluded from that? . . . environmental issues could not exist in a vacuum (author's interview).

Wirth approached Stavins shortly thereafter, not vice-versa. Stavins is even more blunt about the legitimation process that was central to getting new ideas accepted:

> I think it's easy to forget that because the political landscape has changed so tremendously in regard to [market] instruments in the intervening years. . . . We had to be careful about how we approached this. After all, economists had been pushing these ideas for 30 years and the political process had been ignoring them. So it wasn't enough to just present the ideas. It's not the ideas that mattered. What really mattered was the framing, the packaging. . . . The most important article in the newspapers that led to the breakthrough of getting attention was by [columnist] Peter Passell in *The New York Times*. . . . [Passell wrote that] it's not so much what it says, but who said it. And when [Passell wrote that] who said it was important, he wasn't referring to Stavins, he was referring of course to two senators, a Republican and a Democrat. He makes it very clear in the article and that was what was of critical importance (author's interview; Passell 1988).

Stavins noted that even in the United States, where the compatibility of the market and environmental protection had been pushed throughout the 1980s, the ideas still had to be framed in a way to gain consensus from the environmental and, in the U.S. case, the non-Republican constituency. In *The New York Times* article referred to above, Passell (1988) drove home the importance of Senators Wirth and Heinz's support for Project 88 when he wrote: "Their imprimatur confers a new political legitimacy on economists' ways of thinking about environmental problems." Similarly, the Brundtland Commission was able to frame issues, and gain publicity for that framing of environment and development, in a way that would find consensus within a very wide audience beyond the elites who interact with the OECD.

Project 88, as initially conceived, aimed at domestic environmental policy, and was not much influenced by the Brundtland Commission report. Nonetheless, because it influenced U.S. policy through the Clean Air Act amendments of 1990 (through tradeable permit schemes for sulfur dioxide emissions, for example), it helped to provide broad-based legitimacy for a more economistic view of international environmental policy. Wirth believes that it made the Clean Air Act amendments possible. "Whether it was Project 88 that did it, it's certain that Project 88 legitimized a lot of the things the Bush administration was trying to do" (author's interview). This legitimation carried over into U.S. foreign environmental policy in the lead-up to and, to an even greater degree, following Rio, when Timothy Wirth became the U.S. Under Secretary of State for Global Affairs, responsible for global environmental policy in the Clinton administration. According to Wirth, many of the ideas in Project 88 found their way into the U.S. Environmental Protection Agency, Department of Interior, and the State Department prior to Rio and informed the positions taken there (author's interview).

At the international level, the synergy between ideas in Project 88 and ideas of sustainable development as they had begun to be understood following Brundtland facilitated the building of consensus toward Rio to overcome the North-South divide. In Wirth's eyes, Project 88 is "absolutely" compatible with sustainable development. "Sustainable development all has to do with the attempt to link environment and economics. Project 88, I hope, contributes to an understanding of how you make sustainable development work" (author's interview). Such work contributed to allowing the United States, European, and other OECD countries to go into Rio and, despite suspicions of developing country motives and the development side of the Rio agenda, agree on a basic set of governing norms with the South. Agreement was made possible because sustainable development could mean that a liberal economic order and environmental protection could be perceived as compatible.

Although projects such as Stavins' and David Pearce's in the United Kingdom were repeated much more in the North than in the South, the general normative changes in the international economic order meant such ideas could more easily find synergies with the growing domestic consensus among states in the South as well. These changes also acted to disrupt the unifying "developing country" identity of the South and thus their opposition to the liberal economic order as unjust.

In addition to finding synergies with growing domestic consensus among major states in the North and South, sustainable development ideas found support within other UN institutions previously reluctant to incorporate environmental concerns. For example, I have already described the World Bank reforms and its leading role in the GEF. Recall that the Bank's influential 1992 World Development Report on environment and development, like WCED, argued that massive economic growth (3.5 times increase by 2030) is necessary for achieving other ends, including environmental protection and poverty reduction (World Bank 1992b). Brundtland made it possible for the World Bank to proclaim its new "green" image and still promote market liberalization, private property rights, and market-based instruments to change environmentally damaging behavior.

The socio-evolutionary explanation does not provide the immediate causes of the Bank's newfound attention to the environment, which evidence suggests stemmed also from environmental groups' pressure and the leverage they gained through U.S. Congressional hearings in the mid-1980s on the environmental impacts of the Bank's lending (Rich 1994:136–38). Whereas WCED did not cause these changes in the Bank in a mechanical sense, it played an important normative or enabling role. The Brundtland report legitimated a form of international governance consistent with the Bank's development philosophy of export-led growth, open markets, and domestic liberalization, while at the same time it provided an opportunity for a response to environmental criticisms of its lending policies. WCED allowed "win-win" strategies, whereby "links between efficient income growth and the environment need to be aggressively exploited" (World Bank 1992b: iii).

Meanwhile, norms at levels one and two continued to militate against the institutionalization of ideas that challenged state sovereignty from above. For example, chapter 3 noted that global management schemes such as those supported by the Common Heritage norm have fallen out of favor. Indeed, the delegitimating of the Common Heritage Principle (CHP) and the acceptance of its reinterpretation in the 1994 Law of the Sea agreement owed much to social fitness. UN Secretary-General Javier Perez de Cuellar noted as much when he launched negotiations to overcome the impasse in Part XI (the provisions on deep seabed mining) of UNCLOS III that had scuttled the 1982 agreement. The report of the Secretary-General on the new agreement notes that de Cuellar felt that cooperation had become possible because, among other reasons, the "general economic climate had been transformed as a result of the changing perception with respect to the roles of the public and private sectors. There was a discernible shift towards a more market-oriented economy."[33] This change in social purposes among many states meant new common ground could be forged on the meaning of the CHP to reinterpret it to be in conformity with market norms. Taken as a whole, the final outcome of UNCLOS III legitimated sovereign control and market principles, thus it reproduced established norms at level two and three of social structure, much as did the agreements at UNCED.[34]

Other challenges from above have come in the form of proposals to grant more authority to transnational institutions for science or support a more general social movement to increase democratization and common culture at the global level. Indeed, many strands of environmental thought have long supported the creation of a more cosmopolitan world order that limits the role of sovereign nation-states. "In such a world," Dan Deudney (1993:301) speculates, "the nation-state system [would be] pushed somewhat from the center of world political order."[35] The early impetuses behind global environmentalism, as articulated in even mainstream publications such as *Only One World*, were to push in the direction of a greater sense of planetary citizenry and global stewardship (Ward and Dubos 1972). Instead, as I showed in earlier chapters, actual responses have been consistent with sovereign authority and in opposition to global management (except by sovereign states) or relinquishing control to institutions of science. (Note, norms at level one and two militate against relinquishing of sovereign authority, not of state autonomy, which by definition is relinquished to varying degrees whenever a state enters into an international agreement). As others have pointed out, most of the institution-building in response to global environmental concerns has occurred within the confines of traditional sovereign-state diplomacy (Litfin 1993; Conca 1993). The continued reinforcement of sovereign control is in spite of the observation by many scholars and environmentalists that the state is not the appropriate site for, or source of, effective management of environmental problems.

Furthermore, international programs and policies aimed at tackling specific global environmental problems tend to act through states rather than directly on domestic actors or processes. To take one particularly important example, the Global Environment Facility—the primary channel for multilateral lending and aid for environmental protection—reflects the intergovernmental institutional setting of its three backing agencies, the World Bank, UNEP, and the UNDP. As described in one study:

> Each of these organizations was formed through interstate negotiations, and state representatives and interests play a dominant role in their activities. As such, it is the map of states more than any other map that is the spatial frame of reference for decisionmaking by these organizations. This particular state-based worldview is necessarily reflected in the operation of the GEF: the GEF's fourth "operational principle" explicitly states that "the GEF will fund projects that are *country*-driven and based on *national* priorities designed to support sustainable development, as identified within the context of national programs" (Shafer and Murphy 1998:258).

At the same time, Shafer and Murphy note that the environmental problems that GEF funds aim to address cannot usually be defined by state boundaries. For example, biodiversity "hot spots" or regions most targeted for special need of multilateral assistance and urgent attention often do not fall within the boundaries of single states. Like major waterways (such as the Nile or Mekong river basins), these transborder or international areas that fall within the scope of the GEF, as is the case with global issues such as climate change and ozone depletion, are not likely to be best protected by relying on "national priorities" (Shafer and Murphy 1998). According to the explanation presented here, the social structure of the international system reflects deeply institutionalized norms at levels one and two that explain this sovereign state-territorial framing of responses to global environmental problems that cannot be easily overcome.

Hence, it is not surprising that UNCED reproduced norms consistent with the practices of sovereign statehood and control. Although various ideas about environmental governance pose challenges to sovereign state control, attempts to institutionalize such changes have generally failed because of the perception that they threaten the identity and status of some states. Levels one and two of social structure tend to enable only a limited range of responses to global environmental problems.

UNCED OUTCOMES AND LIBERAL ENVIRONMENTALISM

Were the UNCED outcomes themselves completely preordained by the legitimation of liberal environmental norms? The answer is of course no. Any given set of international negotiations includes competition among specific interests, complex dynamics, and unexpected and nonpredictable events. Nonetheless, the UNCED negotiations clearly did occur in the context of a relatively accepted set of legitimating norms around the concept of "sustainable development" and, I have argued, its legitimation as liberal environmentalism. Much of the legitimation had already occurred within the United Nations system and had been reinforced through multilateral fora such as those listed above and in chapters 2 and 3. The micro processes through which this happens have been described by others with different terminology—Ernst and Peter Haas (1995) describe it as learning, Gunnar Sjöstedt (1994) describes it as building consensual knowledge, and Oran Young (1996), on a slightly more macro level, describes it as institutional linkages. Nonetheless, the underlying phenomenon, I would argue, is the same as coming to use a common, and legitimate set of norms on which governance is built.

At the level of process, the UN system itself reflects existing normative compromises in the international system. The various organizations and actors reproduce that normative structure in their activities, which, after all, depend on support from their state sponsors. Indeed, their legitimacy depends on it. As a UN process, UNCED undoubtedly reflected norms that had made headway within various relevant institutional arrangements within the UN system, and helped to create an environment for the normative compromises produced at Rio. Sjöstedt (1994:82), for example, in looking specifically at the UNCED process, focuses on how what he terms the "UN bureaucratic-organizational culture" colored UNCED consensual knowledge. He notes that activities of relevant functional agencies (for example, the World Bank and UNDP for poverty reduction or the FAO for the relationship between agricultural productivity and the environment) had direct input into decisionmaking and framing of problems. Moreover, the UNCED secretariat called on the expertise within those agencies for issue clarification. As the process unfolded, the relevant agencies often helped fit national reports or particular problems into a conceptual framework and program language understandable within and across UN agencies involved. This institutional culture did not determine the content of particular policies outright. However, observing that culture gives some indication at the level of process of how the existing social structure of international politics gets reproduced on the micro-level, even when it evolves in novel ways such as in the framework of Agenda 21—a creation of the UNCED secretariat (Sjöstedt 1994).

Similarly, learning that had gone on within states and the UN system about sustainable development made the more radical proposals of the

South Centre, for example, seem almost anachronistic or a remnant of the NIEO period. Neither did the South Centre (1991) offer up a coherent alternative to "sustainable development." Hence, the proposals reminiscent of the NIEO did not muster the unified support necessary to achieve success at Rio. At the same time, norms such as the entrenchment of sovereign control of resources and even more radical proposals, such as a right to development, gained acceptance because they could be framed as compatible with the norms of liberal environmentalism. Meanwhile, proposals to apply the common heritage norm to biodiversity faced strong resistance. It should not be so surprising, then, that even the United States could not succeed in straying from this consensus with, for example, its position that trade measures to protect the environment should be allowed under certain circumstances (Kovar 1993). Liberal environmentalism, to maintain legitimacy, had to support an open international trading system.

Overarching this entire process are dominant norms of international society that the UN system and its components both reflect and reproduce. The ideas around sustainable development set the path to Rio, but their interaction with the evolving international social structure gives the broader picture of how environmental governance changed to institutionalize liberal environmentalism.

CONCLUSIONS

This chapter has argued that a socio-evolutionary approach is more enlightening than an epistemic communities approach to understand why international environmental governance evolved as it did since 1972. It showed that economic ideas were the source of many of the major developments in the normative evolution of environmental governance, but new norms did not simply arise as a result of those ideas, whether or not they were carried by a specific community of experts. Instead, ideas did or did not become institutionalized as governing norms based on their interaction with the existing social structure of international society.

Ecodevelopment failed whereas sustainable development succeeded because ideas around the latter found legitimacy in key policymaking institutions such as the OECD. However, that legitimacy was not enough. Policy entrepreneurs had to use the legitimacy of the OECD to promote those ideas. When they successfully linked those ideas to the Brundtland Commission process, it provided added legitimacy to a wider government and nongovernmental constituency in the North and South. Those ideas, although many had roots in earlier thinking about environment and development, successfully broke from previous thinking by reframing environmental protection in the language of economic growth. The fitness and interaction with the wider international social structure helped to select a particular pathway from the Brundtland Commission report to liberal environmentalism. In doing so, it entrenched the most significant shift in global environmental policy since 1972—the shift from considering the environment mainly in the context of environmental protection alone to governing norms that now link virtually all global environmental action with liberal economic norms that promote growth.

Disentangling the causal chain of the three factors of ideational success is not easy. For example, UNCED not only reflected an emerging consensus on the proper norms for the international political economy, but may also have played a role in legitimating those norms. The fact that social structure is constituted by practices of actors makes such linear causal thinking inappropriate. Nonetheless, the conditions of entrance for new ideas and norms does suggest causal weight can be attached to the three factors identified in combination—legitimacy of new ideas, fitness with social structure, and fitness with social purposes or identities of major states—and that they reveal a process through which new norms get selected. Since UNCED also promoted some new norms not yet well institutionalized in international social structure, the question remains whether these new challenges will seriously contest existing norms. For example, the increased activity and legitimacy of non-state actors might find openings to gain further legitimacy. Contestation of norms does not cease once they become institutionalized; rather the interaction of practices of actors and the social structure those practices constitute is an enduring condition of world politics.

Theoretically the chapter has attempted to move away from strictly rationalist conceptions of the influence of ideas on international governance to answer some of the critiques raised in the ideas literature. Since the explanation outperformed an epistemic communities explanation in what could be considered its paradigmatic case, it suggests some confidence in its utility for explaining normative evolution in other cases as well. At the least, the socio-evolutionary approach suggests one way to think systematically about the interaction of ideas and their environment, and thus merits further research and refinement. More importantly, the questions it addresses are critical for those interested in understanding the type of world order that institutions promote.

Chapter 6
CONCLUSION

THIS BOOK HAS put forward two major arguments. First, the advent of sustainable development thinking ushered in the institutionalization of liberal environmentalism. This form of international governance predicates environmental protection on the promotion and maintenance of a liberal economic order. Under liberal environmentalism, a liberal international economic order, privatization of global commons, and market norms are not only perceived as compatible with environmental protection, but also necessary for successful incorporation of concern for the environment in the practices of relevant state and non-state actors.

The concept of liberal environmentalism owes some intellectual debt to John Ruggie's concept of "embedded liberalism." Ruggie (1983), drawing on the work of Karl Polanyi, uses that concept to argue that the post–World War II multilateral liberal economic order was predicated upon domestic intervention. In other words, the architects of the liberal order explicitly designed institutions that allowed governments to intervene in their economies to cushion the effects of free trade and financial flows. My position is similar to Ruggie's to the degree that I focus on the social structural basis of a liberal economic order as embodied in legitimating norms. In Ruggie's case, the legitimating norms of international liberalism included, and arguably required, support for domestic intervention. However, I do not argue that liberal environmentalism is a compromise necessary for the legitimation of the liberal economic order promoted since the end of the Cold War, at least not at present. Instead, I take the more modest position that the legitimation of environmental concerns in the international political economy has involved a process of introducing ideas about the environment that, to gain legitimacy, required some compatibility with the kind of economic order dominant at any given time. Environmentalism has not yet become a central pillar of the international political economy; it competes with a variety of social purposes in the construction of the international economic order. Nonetheless, I have also argued that the growing importance and prominence of environmental concerns in global governance owes much to its formulation in norms of liberal environmentalism. These enabling and constraining implications of liberal environmentalism for global responses to environmental problems will be one focus of this concluding chapter.

In at least one respect, however, this book has moved beyond Ruggie's arguments: it introduced a new approach to explaining the processes through which particular forms of governance become institutionalized. Rather than looking mainly to domestic compromises in dominant states to explain the construction of international governance structures, as Ruggie does, I have put forward an explanation that begins with a recognition that international social structure is an evolving set of practices in which new ideas and forms of governance must make headway to gain legitimacy. From this starting point, I developed the second main argument of this book, that the evolution of international environmental governance could be best explained by a socio-evolutionary approach.

This explanation attempts to push forward the literature on the causal role of ideas by showing a way to understand the interaction of ideas (and the norms they support) with the social structure they encounter. The focus on social structure draws attention to the context in which state and other key actors attempt to build governance structures to address global problems. Such structures, I have argued, are not simply responses to material interests. Rather, a theory based on social structure endogenizes an important source of interests in that identities and interests stem, at least in part, from the social structure in which actors participate. Furthermore, the historical or evolutionary aspect of the approach is meant to reveal that social structure is not just a closed system, but evolves in response to new ideas. Social structure is thus historically contingent, and the socio-evolutionary approach attempts to capture the dynamic way in which governance structures evolve in response to new global problems, but always in the context of an existing normative environment.

Below, I discuss the implications and limitations of these findings for the theoretical understanding of global environmental politics, for international relations more broadly, and for the policies and practices to address global environmental concerns.

THEORETICAL IMPLICATIONS

The Normative Basis of Governance

The focus on norms in this study turned attention to the content of governance, rather than to an explanation of whether cooperation on particular environmental problems occurred. It thus acted as a corrective to rational cooperation and neoliberal institutional literature, which, for reasons identified in the introduction, tend either to ignore or to treat as irrelevant the content of policies promoted in cooperative arrangements or the question of why particular norms are selected over others to guide policies and practices. I am not advocating the replacement of rationalist studies of international regimes, which might lead, for example, to useful conclusions on the requirements for stable institutional arrangements. But I do aim to promote a more critical evaluation of what *kind* of environmental governance is actually being achieved, a vitally important question for theory and policy.

Although I did not undertake an independent assessment of the effects of liberal environmentalism, identifying this norm-complex is a first and necessary step to allowing such an exercise. Such assessments might more carefully examine the impact of norms on particular policies and critically examine what criteria of evaluation these norms produce. For example, what does a particular norm-complex mean for issues of equity (between generations, rich and poor states, societal groups, and so on), environmental quality, or cost-effectiveness? The rational cooperation literature truncates debate on these fundamental issues at the heart of designing institutions to address global environmental problems.

Ideas, Social Structure, Change, and Contingency

The ideas literature in international relations developed as one attempt to re-introduce these questions back into the study of international institutions and cooperation. When used in conjunction with the rational cooperation literature, however, such studies too often take it for granted that cooperative solutions reflect progress on the problem being addressed. The epistemic communities literature is particularly susceptible to this problem since authors of these studies almost exclusively examine new ideas that they believe will lead to progress. Such studies tend to look at how new knowledge can alter interests to facilitate cooperation, without questioning why some ideas succeed over others. This problem leads some scholars to an overly sanguine view of the ability of new ideas to alter international relations in a positive direction. Thus an assumption prevails that states can relatively easily "learn" to alter their definitions of interest, and thus create or change international institutions in response to their enlightened views. In recognizing that new ideas do not exist in a social vacuum, my focus on social structure supports a less sanguine view of the ability of new ideas to change international relations. Change does occur in international relations in response to new ideas about legitimate behavior, or to new purposes of action such as responses to global environmental problems. However, social structure is seen to powerfully select certain ideas so that change, especially at deeper levels, generally occurs in a slow and evolutionary fashion.

I have not introduced a general theory about the rate of evolutionary change or the conditions for large transformations. Research that focuses on major change in international politics often concerns massive disruptions or "shocks" such as hegemonic war, revolution, or economic upheavals.[1] By focusing on a particular area of governance, especially one that has arisen relatively recently to prominence, I have shown the slow process of evolutionary change that continues to occur between such major upheavals. But more research might fruitfully be done to better specify the conditions under which quicker or slower change might be expected. As a preliminary observation, however, this case suggests that new issue areas often compete against existing social purposes and gain prominence in governance structures in large part by finding a fitness with those structures, although in so doing they also may highlight contradictions in accepted norms, and create new resources on which actors interested in change might draw. Since ideas are based in meaning and intention, and social structure is based in intersubjective understandings, the human potential to alter such structures, even if it is highly constrained, is always present.

This conclusion may frustrate anyone who expects a definitive resolution of questions of determinacy and contingency, a topic briefly addressed in the introduction. Some social scientists take the view that even stochastic (chance or unique) events can be studied using scientific methods (King et al. 1994: 11–12), which means that even a process-oriented or historical-based explanation need not rest on a view of the social world as highly contingent. Even if King et al. are correct in their view of stochastic events, however, it does not follow that history can easily be studied this way. The unfolding of history may be "caused" by multiple interacting factors, but it is virtually impossible to predict how those factors will interact, owing to a variety of reasons including high levels of complexity, the role of accidents, learning and other sources of feedback, the fact that the social world is an open system, and a variety of other factors that limit the ability to predict. These limitations are well-known and frequently articulated by philosophers of social science, and apply equally, and sometimes more than in other social sciences, to the study of international relations (Bernstein et al. 2000:43–53). As my colleagues and I have argued elsewhere, a deterministic view of social science rests on a mistaken analogy between physical and social phenomena (Bernstein et al. 2000). A better analogy is to evolutionary biology, which is explanatory but not predictive, and still considered "scientific." My socio-evolutionary approach is not predictive, but process oriented, and explanatory in a historical sense. It does not follow that because particular events can be explained that causes of those events can be discovered that act in a law-like way.

Thus, I am not arguing for complete contingency. As much as one can identify a social or institutional structure at time T, one can identify the environment in which new ideas must make headway. The environment is relatively certain and knowable *at any given historical juncture*. What is less certain is creativity in formulating ideas, accidents including natural or man-made disasters, and so on. Moreover, social structure at T + 1 may be different than at T as a result of acceptance or institutionalization of new ideas, which mostly occurs in an evolutionary way. As explained in the introduction, the part of evolutionary terminology stressed here is on historical contingency and social fitness, not goodness or progress. As Caporaso (1993:80) explains, "The stress on historical contingency and path-dependent behavior suggests that many different institutional worlds are possible. What we observe at any point in time is not necessarily efficient (compared with what might have been chosen had other historical contingencies intervened)."

What, then, were other possible historical branches in the case of global environmental governance? A few possibilities come to mind. First, it might have been that no environmental ideas developed that fit well with prevailing social structures. The result would have been less international action on environmental problems, and a much lower level of institutionalization of environmental norms that did prevail. Environmental governance, to the degree it existed at all, might have appeared much more inconsistently, and only developed when crises so severe erupted that they prompted radical action or, if action came too late or not at all, environmental disasters. Second, alternative ideas might have arisen that fit with other aspects of social structure. Such ideas might have produced different pressures for change or tensions with existing institutions. Third, historical accidents or stochastic events might have altered underlying social structure in other directions. One could imagine that social structure would have been altered significantly from the multilateral and liberal order prevailing today if the Cold War had turned out differently or, going back earlier, Nazi Germany triumphed in World War II (Ruggie 1983). Less dramatically, had neoliberalism failed to take hold or Keynesian economics not declined so precipitously, one could imagine social structure exerting a different set of pressures on environmental norms, and creating an institutional environment in which alternative environmental norms might have more easily succeeded. Following similar reasoning, except looking forward, some alternative futures are explored in the policy implications section below.

The Causal Role of Ideas

Naturally, a focus on the content of governance turned attention to the "ideas" literature in International Relations scholarship. In the absence of the ideas associated with liberal environmentalism, the form of governance institutionalized at the Earth Summit simply could not have arisen, nor was it likely that environmental concerns would have come to play as prominent role as they have in international governance more broadly. The introductory chapter showed that ideas mattered and could not simply be derived from the material interests of dominant actors or from the material structure of the international system. If not for the introduction of ideas about environmental problems and about how the international community should address those problems, it is unlikely that the environment could have made the headway it has in international discourse and action.

The early attempts at global environmental governance achieved only limited success because ideas had not been developed to bring the North and South together in a way consistent with other trends in international governance. The initial ideas presented at Stockholm by the conference secretariat did attempt to bring disparate interests together, and in some ways provided the basis for environmental governance as it would develop over the next 30 years. But the ideas promoted at Stockholm could not fundamentally alter the basic underlying interests of North and South. Neither did they provide a way to conceive of how environmental governance could avoid a challenge to core aspects of international social structure or the direction of governance that either the North or South viewed as legitimate. The ideas contained in the Brundtland Commission were a breakthrough in that respect. For the first time, a set of ideas successfully reframed environmental concerns in a way that could be compatible with dominant norms in the international social structure. Thus, they were much better able to alter the understandings of interests of major states in the North and South. As international social structure evolved at what I termed "level three"—the level of norms concerned with coordination and collaboration to manage interdependence—to reflect the move away from international Keynesianism and toward the "Washington Consensus" of liberal market norms, the aspects of sustainable development most consistent with such norms gained favor.

In the above story, ideas mattered in that they had to be developed by some group and needed to gain legitimacy in key organizations that could promote them, in this case the OECD and then the Brundtland Commission, which in turn had a basis of legitimacy in the wider community. Ultimately, however, I argued that what made ideas of liberal environmentalism successful was not simply their promotion by legitimate groups, but their fitness with an evolving social structure. It is this interaction of ideas and social structure that the socio-evolutionary approach uncovers.

In my attempt to contribute to the literature on the causal role of ideas in international relations, I differentiated between the rationalist and interpretivist use of ideas and placed my approach in the latter camp. However, I argued that a socio-evolutionary explanation could still identify causal factors that lead to the selection of some ideas over others.

Before turning to that approach, I tested an epistemic communities explanation of how environmental ideas became institutionalized. I chose this explanation because it contains within it an argument about why some ideas in particular mattered, that is, ideas that had legitimacy rooted in an expert group privileged by its cause-effect knowledge and driven by principled beliefs based on such knowledge claims. Thus, it appears to offer answers to the two questions left unanswered by rationalist approaches: where do ideas come from and why do they get selected? In this case, the hypothesis posited that the ideas came from a group of ecological scientists whose ideas were selected because of the legitimacy of their consensual cause-effect knowledge claims. I also chose this explanation because international environmental governance is a crucial case for the hypothesis, which makes the findings here of more general relevance for evaluating the usefulness of the approach. An epistemic communities explanation should have performed best in explaining the content of governance in an issue area such as the environment, characterized by complexity and uncertainty and that requires technical expertise to both understand the problem and to formulate solutions.

Despite the promising attempt to bridge the rationalist/interpretivist divide, chapter 4 found that the hypothesis failed in key respects to account for the evolution of international environmental governance or even to identify the process of scientific influence on international environmental activities or agreements. This negative finding has implications both for theory, discussed below, and for understanding the actual way in which scientific knowledge did or did not influence environmental governance (discussed in a subsequent section).

First, I took issue with the assumption of the hypothesis that the causal knowledge of the community informs its principled beliefs. In regard to environmental governance specifically, Haas (1996:27–28) uses this assumption to argue that a community of scientific ecologists "sought to develop social laws from their understanding of the laws of nature." The evidence does not support such a position since "social laws" could not be easily derived from the cause-effect research undertaken by the group, nor do most scientists appear willing to support such a linkage. If anything, the history of ecological science shows that strong debates persist about the proper focus and methods for research and the relationship between research and environmental policy. Indeed, I found an uneasy relationship between scientific research and the environmental values attributed to an ideal-type scientific ecology community.

My findings also challenge a related implication of the approach: that policy choices can, and ought, to stem primarily from objective science. This underlying orientation of the epistemic communities approach is revealed in

Haas's argument "that science is essential for the understanding of global environmental problems, thus shifting the determination of the scope of allocative decisions to the international institutions for science" (1996:1). This statement cannot be sustained empirically in the case of environmental governance. The more subtle theoretical point, though, is that the epistemic communities literature is biased toward finding ways to increase the influence of science on policy since the literature makes the assumption that such policies would best reflect the "objective truth" of the situation, to the best understanding of the time. The link between scientific research and policy proposals requires a more critical analysis to unpack that relationship. In addition, while scientists themselves are often concerned about their social responsibility and informing policy to the best of their ability, many of those most active in global change research also appear to recognize the political and social nature of choices. Whether this is true of other issue areas, the relationship between the truth-claims of an expert group and their policy activities ought to be made explicit, not assumed.

Finally, the individuals most directly involved in communicating scientific knowledge to policymakers often do not fit a strict definition of an epistemic community. A number of studies on international environmental issues have shown that primary researchers are not the main source of scientific advice to policymakers. These studies identify "knowledge brokers," "policy researchers," or "science managers" as more often serving as intermediaries between those who produce knowledge and those who make policy (Litfin 1994; Timberlake 1989; Boehmer-Christiansen 1994a). While some of these individuals are scientists, others are not. It may be that some are influenced in their value orientation by scientific endeavors; however, it is equally plausible that their influences include their own institutional or bureaucratic settings or personal histories. Regardless, the epistemic communities literature cannot capture the link between science, this wider group, and their influence on policy.

A rejection of the epistemic communities hypothesis does not mean a rejection of agency, however. While acknowledging the importance of agency in the formulation of new ideas, I have not presented a particular theory of agency that privileges particular groups. Whether the source of ideas stems from epistemic communities, social movements, entrepreneurial individuals, or advocacy coalitions, the argument presented in the previous chapter is only that a social structure of institutionalized norms, as the environment with which new ideas interact, is a major factor in selecting how and whether those ideas become institutionalized. In that sense, the view of agency is relatively open, but refuses to privilege epistemic communities over other sources of ideas and action. I have also not proposed a theory of why new ideas arise. Discovery, crisis, and policy failure are three of the more likely contenders, but these factors are beyond the scope of this analysis, which only argues that when new ideas do arise, they interact with existing social structures.

The socio-evolutionary approach attempted to move the discussion away from a focus on an expert group alone, and toward the interaction of ideas with their environment. Because international social structure is constantly evolving in response to the institutionalization of new norms and the altering of old ones, the socio-evolutionary approach lends itself naturally to an interpretivist methodology. The content, in terms of meaning, of social structure must be investigated at any given time as the environment in which new ideas compete.

Nonetheless, causality is evident; and that makes the approach something more than a purely interpretivist endeavor. Factors and causal mechanisms can be identified that make some ideas more likely candidates for institutionalization or legitimation than others. Even though these factors, such as social structure or ideas, are based in understandings of meaning and are historically contingent, they can still possess explanatory weight. I found Ruggie's notion of "narrative causality" useful in contrasting the causal weight of ideas, norms, and institutions with the formal causality characteristic of the physicalist world, and also noted that both causal and constitutive modes of explanation were at work in the socio-evolutionary approach.[2] This approach is explanatory not only in the sense of identifying the social structure and positing its causal weight; it also emphasizes that specific factors can be identified that reveal processes through which these meanings evolve.

For example, I argued that the legitimacy of ideas within a privileged expert group, even when group members disseminate those ideas within bureaucracies, is not sufficient, nor even necessary, for the acceptance of new norms or changes to existing norms. The perceived legitimacy of the carriers of new ideas is important, but can be gained also through the legitimacy of key institutions through which they act. In this case, the OECD in Paris was such an institution in the realm of public policy, and especially economic policy, among its member states. Even then, however, the selection process of new ideas also involves fitness with existing social structures and with the social purposes of dominant states. These factors in combination had causal weight.

Admittedly, the interrelationship between these factors, particularly the last two, leaves me open to the criticism that they are not discrete independent variables in the positivist sense. Nonetheless, I would argue that social structure is a real structure that regulates and constitutes the identities, interests, and behavior of key actors in the international system. So while this approach is not directly testable against a rational choice approach that takes interests as given, it does identify factors that shape international environmental governance and provides a systematic way to explain the process through which some ideas get selected over others.

The process of institutionalization outlined here also contributes to the constructivist research agenda more broadly and might explain ad hoc findings in existing studies. For example, Finnemore's study of UNESCO's attempts to promote transnational scientific research shows that what started as a project to promote norms of free-flowing transnational ideas ran up against an international social structure that switched from "postwar Kantian transnationalism to Cold War Hobbesian nationalism" (1996a:49–52). As a result, norms changed from the promotion of scientists and research to building state capacity. Although she uses different terminology, these findings are perfectly consistent with a socio-evolutionary explanation contingent on changing norms at level two of social structure following the second world war and prevailing norms of "negative sovereignty." The explanation might also usefully be applied to other issue areas such as human rights. For example, explanations for the limited institutionalization of norms of humanitarian intervention, or their framing, might be usefully analyzed in terms of their interaction with changing social structure over the last 50, or even 100, years.

EMPIRICAL AND POLICY IMPLICATIONS

Scientists, Economists, and Environmental Governance

The findings on scientific ideas and scientists suggest their influence works quite differently than the way suggested by the epistemic communities literature. In support of their influential role, I found that environmental or ecological scientists, and scientific knowledge about the environment in general, clearly did influence the rise of global environmentalism. Individuals and groups of scientists often played significant roles in identifying environmental problems and have been called upon to play a variety of roles in governance, including monitoring, assessment, and technical advice. Similarly, some ecological ideas have been taken up in formulating various international environmental policies.

However, chapter 4 directly challenged scientists' primacy in governance in terms of formulating its content, allocating resources, or providing the legitimating basis for institutions that enable and constrain the behavior of major actors. Perhaps most significantly, contrary to the potentially most powerful implication asserted in this literature, scientists are largely excluded from allocative decisionmaking and often eschew such roles.[3]

Chapter 4 also challenged the claim that the basis of the influence scientists have exhibited in environmental governance to date has rested on their consensual knowledge and principled beliefs. Indeed, consensus on environmental problems often came after substantial political responses had already occurred, as in the case of ozone depletion. I found little evidence to support the presence of a strong consensus on values within groups of active scientists, apart from perhaps a support for scientific research itself. When communities did arise to address particular problems, and then pushed policymakers for a response, the kind of action proposed tended to be purposely general in scope (for example, reduce greenhouse gas emissions to limit temperature increases to 0.1 degrees C per decade) and rarely engaged questions of value trade-offs and modalities. When specific policy prescriptions were put forward, they did not possess any particular causal weight over and above other, nonscientific, considerations. With the exception of the Precautionary Principle, few norms could be attributed to specific values associated with scientific research on the environment. Support was not found for the argument that the legitimacy of the current forms of environmental governance stemmed from an epistemic community.

If scientific influence did not work in the way epistemic communities' literature suggests, then how? Although I did not attempt to identify a definitive pathway through which scientific knowledge fed into international environmental activities, a few very broad observations can be gleaned from the evidence. First, the pathway of scientific influence is less linear and predictable than the epistemic communities literature presents. Second, initial influence depends on entrepreneurial scientists or knowledge brokers, who either through their own entrepreneurial efforts, media exposure, or, less frequently, through a concerted bottom up organization of scientific research (e.g., the AGGG), manage to raise the profile of an environmental problem sufficiently to get it on the international agenda. Scientific consensus does not appear to play a privileged causal role at this stage. Moreover, once on the agenda, the political environment shapes policy-relevant research as much as vice-versa, a finding reinforced by recent comparative work on the science-politics nexus in international environmental regimes (Andresen et al. 2000). Finally, despite the centrality of science to an understanding of global environmental problems, scientists and scientific organizations played only limited roles in each of the three norm-articulating events on which I focused, for reasons summarized in chapter 4.

A second set of implications for the role of science relates to the relationship between scientific research and governing structures. One concern relates to the way the literature on the science-policy nexus can bias what an analyst might see in examining environmental governance: a narrow focus on a particular scientific community can lead researchers to assume that policy outcomes that do not reflect the goals of the scientific community are part of an erosion of policy rather than simply an outcome that reflects a different definition or understanding of the policy in question. In this case, a narrow focus on a scientific ecology epistemic community—whose existence as a coherent group I found little evidence in any case—leads to the erroneous conclusion that environmental governance now faces a backlash from rules and principles of trade regimes and market challenges at domestic levels (Haas 1996:43–44). That misses the compromise of liberal economic and environmental norms at the heart of liberal environmentalism. It also misses how policies that might be perceived as external challenges in reality fit with this form of governance. Thus, an epistemic communities approach obscures the actual norm-complex at the heart of international environmental governance and the most significant shifts in that norm-complex over the last three decades.

Chapters 4 and 5 demonstrated that these shifts in environmental governance have themselves affected research and advice on how to address global environmental problems and the reciprocal influence of social structure on the generation of new ideas. Transnational research networks increasingly focus on questions that fit within a liberal environmental framework and governments increasingly have taken control of scientific and technical bodies set up under international agreements, or that feed directly into international agreements, to research or monitor specific environmental problems. For example, chapter 5 described the changes made in the composition and focus of Working Group III of the Intergovernmental Panel on Climate Change. For the second assessment report (1995) the working group included more economists and focused on cost-effective policy responses that fit with research programs consistent with liberal environmentalism.

These changes might even affect research at more basic levels. For example, Donald Worster (1993) suggests that ecology has evolved to be more politically realistic and human centered in line with sustainable development and adaptability to ecosystems. This position might fit better with liberal environmentalism than the ecology of the 1960s and 1970s. However, the ways in which basic research might be affected by these broader social forces requires more in-depth study than has been attempted here and might be better explored within the science policy or history or sociology of science literatures.

Chapter 5 also highlighted the role of economists and economic ideas. While it might be tempting to portray economists as an epistemic community in environmental governance—significant groups of economists active in policy do exhibit a high level of consensus on cause-effect relationships and policy prescriptions[4]—that would be misleading. In general, this group is not a promoter of specific environmental values or an independent force for social change toward a more ecologically based social system.

However, in one respect economists did fit the definition of an epistemic community in that the value system promoted by many economists seems to fit with the basic tenets of the economic theories with which they work. Although I did not survey a wide number of economists, the following comment by Robert Stavins supports the above position. He believes that while many individual economists might be driven to study environmental questions because they find them interesting or care about the environment, the values their work supports likely stem in part from their economic training:

> Economics is obviously value laden. Just the notion of Pareto-efficiency or cost effectiveness or anything else is an expression of values. . . . That you should worry about minimizing costs or maximizing utility for the greatest number, that is obviously a value system. And it's pretty difficult to go to graduate school, do a Ph.D. in economics, and not come out of it with some internalization of that value system. That it makes sense to think of issues as *ceteris paribus*, let's take the goal as given now . . . that inevitably takes one to the notion of market-based instruments for a pragmatic reason. . . . I think it's the attribute of cost-effectiveness and dynamic efficiency . . . that drives economists to do it (author's interview).

Similarly, in a study tracing the strong anti-regulatory stance of American environmental economists, Okke Braadbaart argues that strong, zero-emission, regulatory policies "went against the grain of everything economists stood for. They violated the conviction of many economists that markets offer a superior solution to policy issues than government intervention" (1998:139). Furthermore, the American domestic debate, stemming from academic opposition to the tough regulations of amendments in the early 1970s to the Clean Air and Water Acts, generated much of the academic work that so strongly contrasted regulative- and incentive-based environmental policy.

Despite this underlying set of values, however, I did not find evidence that an identifiable transnational network of economists acted as a community to push environmental governance in a specific direction. Rather, by virtue of their legitimate positions in key institutions and public policy-making generally, governments have called upon economists to formulate policy responses in line with their professional work. For example, Stavins, the lead researcher of Project 88 in the United States, said he had little or no contact with similar research programs in other countries, or in multilateral institutions such as the OECD or the World Bank, until well after his project got underway. Only then did policymakers call on his expertise to help formulate international policy (author's interview). In other words, while these economic ideas existed in the profession, it was not a group of economists driven by a concern with the environment who were the main cause of the shift in international norms.

Only recently, long after many norms of liberal environmentalism have appeared in international environmental agreements and practices, are economists attempting to coordinate their activities to promote political action. Their interaction in policy exercises promoted by governments seems to have brought a number of interested economists together. For example, in 1997 more than 2,000 economists issued a joint statement (and released it at a press briefing in Washington, D.C.) that the United States would be able to reduce its industrial emissions of greenhouse gases to slow global climate change in a way that would not damage the economy (Reuters 1997). The thrust of the statement, written by five leading economists and signed by about 2,000 others, was that well-designed policies relying on market mechanisms "may in fact improve U.S. productivity in the longer run." The statement explicitly endorsed a system of market mechanisms, such as carbon taxes or trading of marketable emissions permits among countries. What is remarkable about this event is not the position taken, which fits very well with widely accepted views of environmental economists, but that three of the five economists who wrote the statement—Kenneth Arrow, Dale Jorgenson, and William Nordhaus—served as authors or advisers to Working Group III of the IPCC at various stages of the process. (Paul Krugman and Robert Solow were the other two main authors. All five are highly respected in the economics community and Arrow

and Solow are Nobel Prize winners.) Again, the reciprocal influence of social structure and research seems to be at work. Just as economic ideas have
influenced environmental governance, so too has involvement in activities
related to international environmental governance influenced changes
within the economics profession and its work.

The success of economic ideas suggests that ideas that do receive attention depend on their ability to make headway in key policymaking institutions. In addition, ideas have to be able to generate coalitions of like-minded actors in decision-making roles. Ecodevelopment, for example, could
not achieve what sustainable development did. These last two points are
interrelated in that the fit with institutional norms and broader social
structure and the legitimacy of institutions that carry ideas made a difference when weighing the impact of ideas generated by the OECD as compared to IUCN and UNEP.

More work might be fruitfully done on the differential power and legitimacy of various international organizations and networks. In the case of
environmental governance, I found, for example, that the OECD played an
extremely influential role, at least in the late 1970s and 1980s. Few studies
have examined specifically the important role the OECD in Paris plays in
international governance as a source of policy ideas and influence.[5] While a
number of analysts note the power of the IMF and World Bank, for example, which have direct financial levers on governments, the more subtle influence of organizations such as the OECD and Trilateral Commission deserve more attention.

Further research might also usefully examine how international norms
are transmitted to the domestic level or across a wide range of actors.
Here, I assumed that such influence occurs, but stopped after identifying
the norm-complex among international institutions and practices of
major actors in their interactions at the international level. Some recent
research that stems from comparative politics and transnational relations
has begun to take up the question of how norms are then transmitted.
For example, Martha Finnemore's works on how international organizations can act to "teach" norms to governments, and Kathryn Sikkink's
work on issue networks and advocacy coalitions, propose promising avenues for further research on how ideas and norms might move from the
international to the domestic level or across states.[6] Similarly Thomas
Risse-Kappen's (1995) work on transnational relations more broadly has
attempted specifically to address under what conditions networks of actors can carry ideas across various levels of governance, and his work with
Stephen Ropp and Kathryn Sikkink (1999) addresses how international

human rights norms are implemented domestically and affect political transformation processes.

The Compromise of Liberal Environmentalism

This book has made much of the influence of specific sets of ideas on the evolution of environmental governance. It might be objected, though, that the reason liberal environmentalism gained prominence is simply that the policies it promotes perform better in achieving environmental goals. In other words, liberal environmentalism is a rational response to domestic policy failures of the 1960s and 1970s or their inability to generate international action. Some of the comparative environmental policy literature hints at this position. For example, Weale (1992) argues that the poor performance of expensive regulatory policies in a number of Western states led to the search for alternatives.

Such a position is unsatisfactory, however, for two reasons. First, the perceived failure of one set of policies does not then determine what will replace it. The introduction and acceptance of new ideas still requires explanation. This is especially true in terms of the timing of the acceptance of new ideas. As chapter 5 emphasized, ideas associated with liberal environmentalism had been around at least since the late 1960s, yet gained prominence decades later. Their acceptance cannot thus simply stem from their inherent "truth" or come from being "good" ideas.[7]

If one looked only at the range of economic ideas available, a set of ideas associated with a "green" international political economy seemed a more obvious direction toward which international environmental governance might have steered. Recall Eric Helleiner's identification of a distinct and relatively well-developed set of economic ideas that pose an alternative to liberal environmentalism and that have varying levels of support among environmentalists and ecological economists. Some of these ideas fit with liberal environmentalism, while others are radically different. For example, according to Helleiner (1996), a "green" political economy shares with what I call liberal environmentalism a distrust of statist economic planning and encourages small-scale markets. However, unlike liberal environmentalism, a "green" political economy strongly opposes large-scale rational and global economic integration along free market lines (Helleiner 1996:70). In contrast, liberal environmentalism takes a view consistent with Helleiner's description of liberal international political economy theories, that "environmental problems are caused primarily by imperfectly functioning markets and inadequate regulatory frameworks, problems which [liberals] think it is possible to remedy through alternative pricing mechanisms and institutional reforms."[8] Interestingly, many of the ideas Helleiner identifies fit much more closely with the more radical proposals of ecodevelopment, which in practice have largely been pushed to the margins.

Second, if liberal environmentalism were simply a rational response to earlier policy failures, one would expect clear signs that theories and policies associated with the new approach will outperform the policies they are meant to replace. A hypothetical comparison might then be drawn between the rise of liberal environmentalism and the rise of Keynesian economics. For example, Albert O. Hirschman (1989), citing classic research on how Keynesian economics came to the United States, suggests those ideas provided a response to the protracted Depression of the 1930s. The apparent ability of Keynes's theory to both predict the economic outcomes of the period and to offer policies in response made them highly persuasive in the United States, where they first gained policy prominence (although a number of political and administrative factors have been put forward that subsequently limited their influence) (Hall 1989a). As Hirschman has put it, "Seldom in history were the basic propositions of an economic theory so strikingly confirmed by events as during the 1938–1945 period in the United States. Shortly thereafter, the ability of government spending to energize the economy and to drive it to full employment . . . was taken as another, more positive demonstration of the correctness of Keynesian analysis."[9] This explanation has some similarities to McNamara's explanation for the rise of neoliberal policies in Europe in the 1980s, cited in chapter 5.[10]

This view does not hold up well in the case of liberal environmentalism, however, because little evidence currently exists for the greater policy effectiveness of ideas associated with it, as I show below. Nonetheless, supporters of liberal environmentalism had one advantage Keynesian economists did not: their ideas fit with the prevailing economic orthodoxy and practices promoted by the most powerful states and international institutions. In this way, it became relatively easy to convince the wider economics and environmental policy communities to pursue liberal environmentalism, even though the evidence to date does not support the position that these ideas work better at achieving environmental policy goals. Empirical research is only beginning to study the relative merits of market instruments, for example, and those studies are inconclusive (OECD 1994a).

Even on efficiency grounds—where the arguments for policies dictated by liberal environmentalism should be strongest—the evidence to date is inconclusive. For example, an OECD report (1994a) recognizes that markets may not always behave as economic theory predicts, implementation of market-friendly environmental policies may be more difficult than assumed, and the politics of environmental policymaking makes the selection of instruments and policy perspectives more complicated than assumed (see also Majone 1989:116–143). Given that the track record of such policies at the domestic level does not demonstrate superior performance over other types of policies, there is little reason to believe results will differ at the international level or that they deserve promotion over other approaches by international institutions. In explaining why such policies might be chosen regardless, Majone argues that because policy instruments are rarely ideologically neutral, their selection often depends on factors other than their effectiveness:

[W]hether one prefers administrative measures or economic incentives to control pollution seems to depend at least as much on philosophy and ideology as on the technical properties of the two approaches. Those who favor the extension of market principles to previously nonpriced resources like air and water in the name of efficiency naturally prefer market-oriented regulatory instruments, while those who oppose the encroachment of utilitarian principles in social life tend to oppose them (1989:117).

Majone also demonstrates the difficulty in comparing various approaches to combating pollution along any set of consistent criteria, since they are conceptually so different.

Similarly, the OECD study cited above reports that, "Non-economic instruments may work equally well or even better than economic incentives ... since the efficiency and effectiveness arguments associated with economic instruments are not always applicable, as a review of the history of environmental policy instruments discloses" (1994a:35). It concludes that probably a "cocktail" of economic incentives and regulatory measures is the best option. "Economic incentives appear to operate best in combination with, or in support of, other instruments such as direct regulation. Economic incentives alone will not effectively and/or efficiently deal with environmental problems, whether national or international ones" (OECD 1994a:48). Braadbaart's (1998) survey of research on regulative versus incentive-based instruments, focusing especially on the European experience, reinforces these findings.

The 1996 report of the UN Commission on Sustainable Development echoes these views, stating that, "Far too little evidence is available on the practical achievements of economic instruments . . . [to know] whether they live up to expectations." Yet the thrust of the report still endorses their use and norms more broadly reflective of the Earth Summit outcomes (UNSCD 1996). Academic studies of implementation at the domestic level have also begun to question the practicality of incentive-based policies, their negative consequences for distribution of costs of environmental protection, whether they can even be separated in practice from regulatory approaches, and the stark difference between theory and practice in the projected effectiveness of such instruments (Braadbaart 1998; Reitan 1998). At the international level, even former supporters of mechanisms such as emission trading for climate change are beginning to question whether the practical application of these tools will prove effective or equitable in combating climate change.[11]

My argument has been that despite these ambiguous findings, liberal environmentalism still pushes for market-based policies over other possible alternatives. For example, internationally, the OECD promotes economic instruments over regulatory instruments in the implementation of international trade agreements and continues to devote environmental research to issues such as cost-benefit analysis and economic instruments. In addition, international organizations such as the World Bank continue to promote such policies in the developing world, focusing on proper pricing and privatization, although there are some signs that this emphasis is changing. For example, senior Bank officials have said the 2002 World Development Report, planned to coincide with the Rio +10 conference, will reflect a shift from the assumption in the 1992 World Development Report that all development policies and programs could be "win-win." The 2002 report will acknowledge that many of the Bank's promises have not materialized, and thus the need for trade-offs as well as synergies.[12] Such a change, if it occurs, may coincide with hints of broader pressures for change in international social structure discussed below. To date, however, the implication of liberal environmentalism has been that the criteria upon which environmental institutions are evaluated has turned more toward economic efficiency and sustaining the liberal economic order than pollution abatement or environmental quality, and alternative options are not being adequately explored.

To take the most prominent current focus of attempts at environmental governance, climate change has been subjected to analysis along liberal environmental lines perhaps more than any other global issue, and, as shown earlier, these analyses have strongly influenced research and action at the international and domestic levels. Leading up to and since Kyoto, OECD studies, for example, supported this position, and have suggested that deciding on proper policy instruments for the abatement of greenhouse gases should be based primarily on economic efficiency criteria. The main choice one major study presents is between two market mechanisms, a global carbon tax and tradeable emission quotas (OECD 1995:9). Such studies tend to gloss over questions of what criteria are to be used when evaluating costs and benefits. For example, given high levels of uncertainty as to the effects of climate change and a number of external variables related to economic performance in different parts of the world, many assumptions must be made about what the impact of various policies will be. Ultimately, the choice of normative criteria, such as whether and how to weigh costs and benefits to future generations or across regions must be incorporated. Such considerations can change calculations radically (Howarth and Monahan 1996). It still may be possible for economists to incorporate such criteria into cost-benefit analysis, but it may also be that liberal environmentalism limits debate on such issues or leaves such choices to economists who are empowered by the legitimacy given to market principles and neoclassical economic analyses.[13]

A danger also exists under liberal environmentalism that a radical free market position could gain legitimacy, although liberal environmentalism as I have described it does not go to this free market extreme. "Free market" environmentalism eschews any attempts to incorporate social or environmental costs or discount rates for the future as too intrusive and likely to lead to perverse results.[14] In fact, some adherents to the norms I have grouped under liberal environmentalism do support government intervention or international management to correct market imperfections or build environmental markets.

Nonetheless, a related problem arises because the advice of environmental and ecological economists is only being partially heeded. Liberal environmentalism tends to support arguments for creating markets, property rights, deregulation, and an end to subsidies. However, the norm-complex has yet to embrace the more radical proposals that might have the largest payoffs for the environment, such as changing accounting practices, large-scale shifts to environmental taxation, or truly integrating environmental considerations into conceptions of social welfare. Major actors view such proposals as too intrusive to free enterprise and the smooth operation of the international liberal economy, or politically unrealistic.[15] Yet, even leading proponents of market mechanisms and an economistic approach recognize that other goals for environmental policy might be important. Hahn and Stavins, for example, put it this way:

In the economist's version of public-policy heaven, the objectives for policy will typically be efficiency (maximizing net benefits) or cost-effectiveness (choosing the least costly method for achieving a goal). Efficiency and cost-effectiveness however, are by no means the only possible criteria for judging environmental policies. Other considerations might include overall effectiveness, ease of implementation, equity, information requirements, monitoring and enforcement capability, political feasibility, and clarity to the general public (1992:464).

Deeper critiques of the implications of liberal environmentalism are also present in the literature. For example, Chatterjee and Finger argue that the type of environmentalism promoted at UNCED left unexamined the industrial processes and unsustainable economic models that caused the current environmental crises. They view the outcome of the Earth Summit as follows:

UNCED has promoted business and industry, rehabilitated nation-states as relevant agents, and eroded the Green movement. We argue that UNCED has boosted precisely the type of industrial development that is destructive for the environment, the planet, and its inhabitants. We see how, as a result of UNCED, the rich will get richer, the poor poorer, while more and more of the planet is destroyed in the process.[16]

While I have not independently assessed the merits of these critiques, the approach to institutions taken here opens up space for the questions they raise, which are obscured by other approaches in the international relations literature.[17] I would argue it is not enough simply to critique the forms of environmentalism of which one does not approve; the way in which they arise and become institutionalized should first be recognized and revealed. Only then can serious debates occur about the possibilities for change, can honest assessments take place about the merits and limitations of various approaches to environmental protection, and can a deeper understanding be achieved of actual social forces at work and their effects.

Implications for the Future of Environmental Governance

The argument and findings presented here have two sets of important implications for the future of environmental governance. First, they suggest the enabling of policies that fit with liberal environmentalism and the facilitation of cooperation on problems amenable to solutions within this norm-complex. Conversely, policies that contradict key norms of liberal environmentalism are more likely to face strong contestation or not even be considered owing to the prevailing norm-complex. Second, the theoretical findings point toward an examination of changes in underlying social structures as potential sources of change in environmental governance, and the importance of analyzing the interaction between policy ideas and social structure. As mentioned earlier, alternative futures can be examined similarly to alternative pasts, by putting forward possible scenarios sensitive to a number of contingencies in the future trajectory of social structure.

Following from the first set of implications, international cooperation on some environmental problems will be easier if solutions can be found that fit within the liberal environmentalism compromise. Thus, the compromise enables action, but action of a certain kind within institutions that do arise. If institutions cannot be constructed within these normative constraints, international action will be more difficult and disjointed.

In the most significant example, the compromise behind the 1997 Kyoto Protocol of the Framework Convention on Climate Change, detailed in chapters 3 and 4, linked commitments by developed countries to quantitative limits or reductions in greenhouse gas emissions to three market mechanisms that involve transferring "credits" for emissions to help countries meet their targets. It is still to early to know if the enabling conditions of a liberal environmentalism norm-complex has sufficiently shaped interests or can overcome a variety of competing domestic constraints playing out in the climate change debate. The argument here is only that these normative conditions provided, and continue to provide, an opportunity for agreement that would have been more difficult under another norm-complex. The irony may be that the kind of agreement enabled, as many critics maintain, may be vastly inadequate to significantly forestall, let alone stop or reverse, current trends in greenhouse gas emissions that lead to climate change. No claim has been made that liberal environmentalism is the optimum solution for effective responses to climate change, only that any cooperative solution on the problem is most likely to be accepted if it fits within the set of norms legitimated within this norm-complex.

Following from the same logic, the evolution of possible management regimes for global environmental problems could also be expected to occur within the opportunities and constraints of liberal environmentalism. The combination of "common concern" discourse and institutional arrangements that acknowledge the responsibility of sovereign actors for good stewardship and access to benefits, but do not make authoritative claims on behalf of a larger community, indicates the possibility of new global institutions that could, for example, take on some functions originally envisaged in the Authority under the original 1982 Law of the Sea (UNCLOS III) agreement. New norms institutionalized since then enable solutions that could avoid the opposition that the Authority initially faced. Indeed, the 1994 implementation agreement for UNCLOS essentially changed the Authority in a direction more compatible with liberal environmentalism (see chap. 5, n. 34).

Looking back at the 1980s and early 1990s, the identification of specific environmental problems that affected the global commons, such as ozone depletion or climate change, and also affected important economic sectors, raised the stakes in contestation over the Common Heritage Principle (CHP). These pressures simultaneously added urgency to the search for alternatives that could be more easily nested within social structure. The difference between any new management scheme and the Authority would be that any new organization's functions would need to demonstrate compatibility with norms of market liberalism and state sovereignty, that is, practices institutionalized at levels two and three of social structure. Thus, an explanation for the ultimate failure of the Authority as an institution vested with authority over sovereign states also highlights opportunities for alternatives.

The climate change case again provides a recent example since it raises the prospect of global management of a commons problem on an unprecedented scale. Although the institutional manifestations of the mechanisms identified earlier to address climate change are still being negotiated, current proposals revolve around norms consistent with PPP, sovereign authority on actions within each state and in decision making in terms of commitments, and freedom of private corporations to choose how to respond to the new markets created. The point is not that such mechanisms are better or worse for the environment, equity, and so on than institutions as envisaged under CHP, but that social structural pressures have shaped environmental governance in this direction.

Conversely, international environmental problems where solutions that fit within liberal environmentalism have evaded negotiators have proven difficult to address cooperatively. As chapter 3 argued, the lack of progress on a global convention on forest protection and use is such a case. As the prospects for a global convention have dimmed, the trend toward certification and labeling of forest products, as a way to internalize environmental costs where regulatory solutions to forest protection have failed, has increased in legitimacy and viability. Such schemes operate in the marketplace, sometimes with government involvement, although usually, as in the case of the most prominent transnational scheme—the Forest Stewardship Council—without.

There are ironies and contradictions within liberal environmentalism as well. Whereas it opens up opportunities for new forms of management, and perhaps for democratic participation as well, it also reinforces the role of the market. Perhaps nowhere is this contradiction greatest than in the Precautionary Principle. On the one hand, as argued in chapter 3, this norm fits easily with the Polluter Pays Principle and the logic of internalizing costs and market norms. However, in practice, the Precautionary Principle politicizes decisions about risk under uncertainty, potentially empowering government regulation over powerful global institutions such as the WTO or some regional trade agreements. Such institutions currently put the onus on governments, under a high burden of scientific proof, to override liberal trade norms. The Precautionary Principle would reverse, or at least modify, the burden of proof from governments who want precaution to guide decisions on allowing products with potentially harmful effects into their markets, to exporters or producers to show their products are safe for human health and the environment.

The agreement in January 2000 on the Cartagena Protocol on Biosafety of the Convention on Biodiversity highlights this tension within the liberal environmentalism norm-complex. The protocol aims to ensure adequate safety in the development, handling, and use of living modified organisms (LMOs) resulting from biotechnology that may have an adverse impact on the environment or human health. Late-hour negotiations to hammer out the final details of the agreement revolved around incorporating the "precautionary approach" or principle to the transfer of modified living organisms, which it eventually did in the preamble and in Article 10.6. (Article 10 contains the operative provisions on decisions of importing countries on LMOs.) Whereas the final document also included language that the protocol and other international agreements (i.e., trade agreements such as the WTO) are to be mutually supportive, it is not to affect the rights and obligations of governments under existing agreements. It also explicitly recognizes core WTO norms such as nondiscrimination. What this means in practice is uncertain. The precautionary approach in the protocol states that a lack of scientific certainty due to insufficient information of the potential adverse effects on biodiversity shall not prevent a Party from taking a decision on LMOs under the Protocol. Meanwhile, the WTO requires "sufficient scientific evidence" to restrict trade for health and safety reasons (under GATT article XX, which arguably also includes exceptions for environmental reasons). Under these circumstances, the Precautionary Principle and liberal trade norms will co-exist uneasily in practice and conflicts over specific LMOs are likely (IISD 2000a).

Even prior to the protocol, high-profile disputes such as that between the EU and the United States and Canada over hormone-modified beef demonstrated the difficulty in reconciling these two principles. In that case, WTO Panel and Appellate Body rulings went against the EU ban on beef because the EU did not conduct a risk assessment. Such an assessment had to bear a "rational" and "objective" relationship to the ban under the 1994 WTO Agreement on Sanitary and Phytosanitary Measures (SPS agreement), which applied in this case.[18] Significantly, however, the Appellate Body also ruled that under the SPS, risk assessments need not be based exclusively on laboratory science under controlled conditions, but also on assessments of risks in human societies as they actually exist. Recall from chapter 4 that this latter view of science fits much better with the Precautionary Principle. Thus, the Appellate Body ruling overturned the Panel on two important grounds: first, it clarified that WTO members could impose higher levels of protection to human health than prevailing international standards as long as such standards were scientifically justified, and, more importantly for this discussion, opened the door to a broader view of science more consistent with the Precautionary Principle—opening up the debate on what burden of scientific proof is sufficient to limit trade.

It remains unclear, however, whether the Biosafety Protocol can tip the balance toward precaution. Given the reality that the WTO dispute panel process can impose binding decisions on parties, while a similar process has yet to be put in place under any existing environmental agreement, decisions on trade issues seem more likely to follow the pattern of findings against environmental limits to trade in spite of the precautionary approach in other agreements. Yet, if the Precautionary Principle remains legitimate, and its further institutionalization in the Biosafety Protocol suggests it will, the potential for transformation of trade norms and practice remains. At the least, continued tension in the legitimate criteria for exceptions to liberal norms will prevail, although it remains to be seen how the WTO will address its relationship to environmental agreements and the trade-environment relationship will evolve.

In at least one sense, however, the Biosafety Protocol and its inclusion of the Precautionary Principle further entrenched liberal environmentalism. It did so by framing the debate over LMOs narrowly as a trade issue, in effect closing off the possibility of wider agreements on rules and procedures governing research, development, and implementation of genetically modified organism technology in isolation from trade concerns.

Whereas a prediction of how these tensions will play out is premature, the institutionalization of liberal environmentalism and unwillingness within organizations such as the WTO to admit of contradictions with norms such as the Precautionary Principle suggests that unilateral government regulation over and above agreed international standards will remain difficult in the short term. Given the difficulty of changing current institutions and structures, nongovernmental groups and entrepreneurial leaders are turning to new institutions such as private regimes or voluntary schemes such as the Global Compact—a UN Secretariat-sponsored scheme that identifies a set of human rights, as well as labor and environmental norms based on existing UN agreements that corporations can sign on to voluntarily (Kell and Ruggie 1999). While this strategy may make some headway within liberal environmentalism, it is also subject to the limits of corporate self-regulation since the UN has no mandate to independently regulate private corporations. At the same time, some civil society groups, frustrated with the limited ability of international institutions to address environmental concerns under the current governing arrangements, have begun to launch more radical forms of opposition to challenge the legitimacy of existing institutions. The WTO protests in 1999 in Seattle and earlier opposition to a proposed Multilateral Agreement on Investment attempted to take advantage of contradictions within current norm-complexes, including liberal environmentalism, that promise the compatibility of liberal markets and goals such as environmental protection. Given that liberal environmentalism and social structure more broadly has legitimated institutions such as the WTO, the argument here suggests that little normative leverage exists to counter the linkage of liberalization and environmental concerns, although to the degree that institutions in practice appear to produce consequences that belie this understanding, new ideas are likely to arise that may yet reveal and take advantage of contradictions in order to push for change.

The second and final set of policy implications for the future concerns where environmental governance is headed. Two sets of forces are at work. First, the internal dialectic in liberal environmentalism, already hinted at, means that contradictions within norm-complexes can be used by actors to push for change. Liberal environmentalism on the one hand empowers states. On the other, its support for market norms means the potential for other actors to gain legitimacy, a tension also reflected in broader social structural changes in the international political economy. Already there is evidence of a greater role for private or hybrid regimes—the latter referring to regulation by bodies with both government and non-government representation—by organizations such as the ISO or the Global Compact (Clapp 1998; Kell and Ruggie 1999). Transnational market mechanisms such as tradeable permit schemes also create the possibility of private markets among or even within transnational firms.

The second set of forces involves broader changes in the international social structure in which liberal environmentalism is nested. Given the uncertainty of such trends, and the complexity of historical forces at work as one moves to the general level of international social structure, I will simply highlight two possible scenarios of broader change in social structure that could result in an altered environment with which ideas to solve future environmental problems will interact.

First, the spread of broader liberal democratic and human rights norms are reinforcing greater demands for accountability and participation by civil society actors in international institutions. It is conceivable that changes in social structure could occur at level one to reinforce these trends. Some authors, for example, suggest that new criteria for state recognition, based on democracy or human rights, for example, denote a diminution in the scope of sovereign authority recognized as legitimate (Chopra and Weiss 1995; Murphy 1996; Biersteker and Weber 1996; Sikkink 1993). Such a shift is significant because it would mean sovereignty as a legitimating principle that defines the status of territorial states no longer served as its own basis of legitimacy, but rather rested on a foundation that required further legitimacy, such as representation of a population or minimum standards of human rights or welfare.[19] If the legitimate basis of state authority shifts, this could alter constraints and opportunities on governance, further empowering non-state actors who base their legitimacy on such norms. At the same time, to the degree such rights are framed in classical liberal terms, they could reinforce corporate freedom of action from state authority. However, such changes in social structure are far from clear, and sovereignty as the legitimate basis of supreme authority and its coupling with the territorial state as the legitimate form of political organization and mode of allocation for exercising that authority remains well institutionalized (Kratochwil 1995:25; Ruggie 1993). Despite challenges at the margins, sovereignty as status remains firmly entrenched as a legitimating principle in most international institutions and current practice since challenges to it still face strong resistance from major states, even those that most staunchly defend the spread of human rights and democracy.

A less fundamental shift may also be occurring at what I termed level three of social structure. Evidence of this can be seen in the intellectual challenge to the "Washington Consensus" by prominent economists who have put forward a modified set of policy prescriptions dubbed the "Post-Washington Consensus" by former World Bank chief economist Joseph Stiglitz. (He left the Bank before the end of his term after speaking out publicly against Bank policies associated with the Washington Consensus, and especially the Bank's and IMF's handling of economic reform in Russian and Eastern Europe. Various state leaders and prominent economists have also criticized these institutions' handling of the Asian financial crisis beginning in 1997.) While not a radical departure from the Washington Consensus, it includes two important modifications. First, whereas the general instruments and goals of a market economy remain, the new prescriptions recognize the importance of sequencing in reform, and the importance of institutions, rules, education, and so on, in making markets function. The second set of modifications is a more significant departure. They acknowledge that development policy may require trade-offs, including those between economic efficiency and growth and other goals such as sustainable development, increased participation in decision making at a variety of societal levels, and greater equity. In other words, development policy that includes noneconomic goals may not always be "win-win" (Stiglitz 1998, 1999). The proposed shift in the Bank's understanding of the requirements of sustainable development mentioned above suggest that these ideas have influenced at least sections of the Bank bureaucracy that deal with environmental and social development issues. Even Bank President James Wolfensohn began in the late 1990s to distance himself and the Bank from the more orthodox policies of the IMF, and Bank publications started to attack the Washington Consensus on social and environmental grounds (Broad and Cavanagh 1999). If the need for trade-offs becomes more explicitly recognized within dominant international financial institutions, reinforced by demands within global civil society for greater accountability and value trade-offs in institutions such as the WTO or any future global investment regime, this will provide a new set of opportunities for global environmental governance. For example, it may possibly move toward a recognition that environmental goals may sometimes require actions that disrupt markets or that cannot be accommodated within existing institutions built primarily on norms that support and reinforce state sovereignty or the growth of liberal markets as an ultimate goal.[20]

Many of these conclusions are speculative, but suggest that an examination of prevailing norms in a given issue area, and the exercise of spinning out scenarios for change in broader social structures, may offer a variety of avenues for understanding the constraints and opportunities for the future of global governance.

CONCLUSION

In the introduction, I posed the question of whether ideas associated with global environmentalism really implied a transformation of international relations and society. In some ways, the advent of liberal environmentalism does suggest a transformation has occurred. Environmental concerns now regularly appear on the agendas of international organizations that engage in a wide variety of practices, from agencies directly concerned with the planet's resources and environmental quality to those whose main focus is trade, development, or even security. Many states' foreign policies also regularly include high-level attention to global environmental concerns and a wide range of non-state actors directly address global environmental problems and the human practices that contribute to, or ameliorate, such problems. Arguably, the rise in global environmental consciousness and activity has also had broader political consequences. For example, global environmentalism is one of many factors that contribute to an increased awareness of interdependence and, therefore, the need for cooperative governance arrangements at the international or global level. It has also contributed to changes in the North-South agenda by highlighting issues of common concern. On the one hand, these understandings have provided new potential sources of leverage for the South owing to the interdependence of biosphere resources that sustain life on the planet. On the other hand, the recognition of a variety of areas of common interest has helped to undermine the confrontational style typical of the period following decolonization up to the end of the Cold War. Finally, liberal environmentalism itself, by framing environmental problems as inexorably linked to economic activity and concerned with similar development goals, has enabled environmental concerns to increasingly move to a central place on the agendas of international organizations and global discourse.

Yet liberal environmentalism has not transformed the international system itself in ways that resemble the initial proposals put forward by the internationally focused environmental movement. Global environmental concern and action within a liberal environmentalism norm-complex has not moved us much closer to a cosmopolitan world order that has pushed nation-states from the center of world politics, as Deudney speculated it could (1993:301). Instead, the nature of global environmentalism has itself been transformed to fit better within the normative structure of international society. New ideas were indeed required to make that transformation possible, and thus to bring environmentalism into the mainstream of international relations, but those ideas interacted with an existing social structure in an evolutionary fashion.

The ultimate legacy of UNCED and liberal environmentalism is uncertain. I have not argued that liberal environmentalism has been a success in solving environmental problems or improving environmental quality. Neither have I argued that environmental policy has in fact been incorporated into economic policies everywhere to the degree promised at Rio. Indeed, as my discussion of the post-Rio assessment in 1997 indicated, progress on many global environmental issues remains limited at best when measured against the goals established in 1992. Rather, this book has made an argument about how the normative basis of international environmental governance has evolved.

In light of the limited achievements since UNCED, this analysis seems especially appropriate given that UNGA member states agreed in 1997 to reaffirm their commitment to Agenda 21 and the principles in the Rio Declaration on Environment and Development. The affirmation of the norms institutionalized at Rio suggests that these norms remain the core of international environmental governance. At the least, understanding the process behind the evolution of these norms and identifying such trends might allow a deeper critical analysis of why specific policies and programs based on that governance structure have not achieved all that was hoped for at Rio. At the most, I have tried to suggest that debates about the most appropriate such norms, and the possibilities of change, ought to be reinvigorated.

CHAPTER 1

1. Sand 1993:378. See also Pallemaerts 1996.
2. The work on international environmental institutions has expanded rapidly since the late 1980s. See Zürn 1998.
3. Levy, Young, and Zürn's 1995 review of the international regimes literature demonstrates this trend. Another review (Hasenclever, Mayer, and Rittberger 1996) notes that the new literature on norms entails a "radical critique" of traditional regime theory and requires attention to ideational or social structural factors.
4. Notably Finnemore 1996a; and Klotz 1995. See also Katzenstein 1996.
5. For example, Young 1994; and Haas, Keohane, and Levy (1993) emphasize institutional conditions for cooperation and a number of major recent research projects focus on regime effectiveness (Sprinz 1999; Wettestad 1999; Young 1999). Meyer et al. 1997 explain the organizational form, not the content, of what they call the "world environmental regime." Zürn's 1998 review of research on international environmental politics primarily focuses on, and promotes, research on regime effectiveness and institutional design, but notes the promise of some new research on transnational relations, ideas, and communicative action. Still, the research he cites, with few exceptions (e.g., Litfin 1994), takes the goals or values of institutions as given, rather than problematizing them in terms of how issues are framed or why some goals are promoted over others.
6. The term "liberal" can cause confusion. For Americans, liberal usually means social democratic. For Europeans it connotes classical or neoclassical economics. In keeping with its primary connotation in the International Po-

litical Economy literature, liberal here primarily means the favoring of "market-oriented public policy to resolve social and political problems" and a rejection of the perspective of Keynesian economics (Falk 1995:563 fn. 3). Moreover, I use the term liberal rather than "market" environmentalism both to differentiate it from "free market" environmentalism (e.g., Anderson and Leal 1991), and in reference to a liberal international order where state sovereignty is infused with classical liberal values such as "negative" rules of abstention (i.e., against unnecessary interference in the exercise of territorial jurisdiction among sovereign territorial states) and clear property rights. These characteristics of state sovereignty remain important structural features of the international system (Jackson 1990).

7. WCED 1987:43. This definition is the most widely cited, although variations appear elsewhere in the report.

8. Brooks 1992. For a list of about 25 definitions, see Pearce et al. 1989:173–185. See also Caldwell (1990:207) and Moffat (1996) for discussions of definitions. The Declaration is reproduced in IDRC 1993.

9. IIASA proposal for a workshop on "Sustainable Development: Principle and Criteria," quoted in Caldwell 1990:207.

10. On "governance without government" in international or global politics, see Rosenau and Czempiel 1992; Finklestein 1995; Young 1995.

11. I use the term norm in a generic sense to include rules, principles, standards, maxims, and so on. The terms may be used interchangeably.

12. In the case of embedded liberalism, Ruggie (1998:84) demonstrates that extreme versions of laissez-faire liberalism pose a greater challenge to postwar economic regimes than many forms of protectionism.

13. Economists tend to favor tradeable pollution permit schemes over other market approaches to pollution control. Such an approach involves the creation of a market where agents can buy and sell "rights" for actual or potential pollution. Under emission trading, "dischargers operate under some multi-source emission limit and trade is allowed in permits adding up to that limit." OECD 1994a:20. Other economic instruments might include charges or taxes or positive economic incentives, and variations on these themes.

14. For example, Hajer 1995; Weale 1992; Pallemaerts 1994, 1996; Chatterjee and Finger 1994; Sachs 1993; Hawkins 1993.

15. This is empirically true, not a causal claim.

16. Goldstein and Keohane (1993) classify ideas in the latter three categories.

17. I do not follow Finnemore's (1996a:22) sharp distinction between ideas, which are held by individuals, and norms, which are intersubjective, because that distinction corresponds to a neo-utilitarian versus constructivist view of ideas, not to analytic categories that can be related to one another. See Ruggie 1998:20. Rather, my argument is that ideas can have normative qualities, but what is important in identifying norms is the degree of institutionalization.

18. Ruggie 1998:20, quoting John Searle.

19. See endnote 5.

20. Gilpin 1987; Keohane 1984; Krasner 1985. Note, Keohane argues that a hegemon is not necessary for regime formation, but introduces no other theory

of regime formation in his major theoretical statement on neoliberal institutionalism.

21. The debate about U.S. hegemony, whether it is declining, and whether power is fungible across issue areas versus whether hegemony can be issue-specific, is voluminous. See Young (1989, 1994) and Paterson (1996:91–113) for discussions about hegemony as it relates to major environmental issues. For the limited purposes here, I simply assume that the United States is the only possible candidate for hegemony, but admit skepticism as to whether one can even reasonably speak of hegemony in relation to international environmental politics.

22. Jordan 1994a. The debate around additionality is discussed in detail in chapter 3.

23. For a critical review of new Gramscian scholarship in International Relations, see Germain and Kenny (1998) and responses from Murphy (1998) and Rupert (1998).

24. Maurice Strong, author's interview, February 1996.

25. Strong describes how the relationship came about as follows: "Schmidheiny … was a leading businessman who himself had a very strong commitment to the concept of sustainable development. I met him in the early period of my role as secretary-general at UNCED. I liked him and I challenged him to take time off from his business and become my senior business advisor. He thought about it. I went home for a weekend with him and his family—he agreed to do it and he did even more than he agreed to do. It proved to be a very fortuitous choice." Author's interview, February 1996.

26. Politicians and bureaucrats who attended did so in an individual capacity, not as representatives of their states.

27. Cracks began to appear in this coalition only after industrialized states agreed to some form of mandatory reductions in greenhouse gas emissions at the 1997 Kyoto, Japan meeting of parties to the 1992 Framework Convention on Climate Change. For example, energy giant Royal Dutch/Shell Group of Companies broke from the GCC in 1998 and pledged to cut emissions of greenhouse gases from its global operations by more than 10% by the year 2002 compared with 1990 levels. British Petroleum (now BP Amoco) made a similar pledge in 1998. See ENS 1998. It appears that major mergers in the industry have not adversely affected these commitments, and both energy giants recently announced internal carbon emission trading markets.

28. Robert Cox's historicism is an exception to the limitations of this literature, because it attempts to address the potential for agency in social movements as sources of counter-hegemony or the basis of alternative hegemonic blocs. However, most applied research in the neo-Gramscian school gives primary causal weight to economic variables, and cannot account for ideational or institutional change independent of economic forces, without importing subsidiary ideational explanations. The socio-evolutionary explanation is not so much a critique of work such as Cox's, as an attempt to take its insights without being limited by Gramscian foundations.

29. Risse-Kappen (1994) makes a similar point.

30. Hajer (1995) presents an "argumentative" approach that attempts to correct

this problem by focusing on how language affects identities and interests of coalitions of actors by creating "discourse coalitions" around particular story lines, as opposed to interests. However, his focus remains primarily on language and less on how and why ecological modernization became institutionalized in the first place.

31. Especially Florini 1996. The approach is also influenced by institutionalist theory as applied to international politics (e.g., March and Olsen 1998; Weber 1994), and the social structural aspects of the explanation draw heavily from Busumtwi-Sam and Bernstein 1997. The term evolution has a long lineage of use and abuse in both the natural and social sciences, the latter being particularly fraught with ideological manipulation. I want to distance myself from the teleological or value-laden use of such theories characteristic of early Social Darwinism. Such approaches posit that social evolution is moving toward a progressive goal, usually western ideals of civilization or social organization. Florini (1996:370) correctly identifies this common problem: "Such applications were based on a fundamental misunderstanding, if not a deliberate misuse, of the basic idea of evolution through natural selection—that 'fitness' is a purely contingent phenomenon. If some individuals or groups prosper while others falter, this means nothing about their relative virtue. It means only that the former happened to have a combination of attributes, resources, and/or luck that better met the environmental demands of the moment than did the latter."

I chose the word evolution because I borrow the concept of "fitness" directly from Darwin's formulation and am attracted to the notion of evolutionary change as a useful analogy for the historical processes my socio-evolutionary approach identifies. However, the borrowing of concepts should not imply an endorsement of any particular variant of evolutionary theory nor do I claim that the socio-evolutionary approach as a whole can be derived from evolutionary theory as studied by biologists or geneticists. On efforts to apply concepts to the study of international relations *directly* analogous to evolutionary mechanisms and derived from specific theories of evolution in biology, see the special edition of *International Studies Quarterly* (1996).

32. The "selection" or "success" of norms in this case means they became the authoritative or legitimate responses to the problem in question, not that they necessarily best achieve some given ends. In this formulation, ends and means are not treated separately (Weber 1994:7). For example, the legitimacy (success) the trade regime bestows on notions such as reciprocity is neither as a means nor an end. Rather, as Kratochwil and Ruggie note, "in a quintessential way, [such norms] *are* the regime—they *are* the principled and shared understandings the regime comprises" (1986:770, emphasis in original).

33. Chapter 5 presents a fuller discussion of the concept of social structure.

34. Weber (1994) uses the concept "social fitness," but not in reference to an evolving social structural environment.

35. Hollis and Smith 1990:49–55. For the purposes of the discussion here, I am glossing over the question of whether this "covering law" approach to posi-

tivism (i.e., the law "covers" the event or situation in question) can really accommodate causal laws, or just regularities. See McKeown 1999:162–164, who
draws especially on Miller 1987, on this point.

36. Ruggie 1995:95. See also Cox 1986.

37. For a discussion of the various reasons why, see Bernstein et al. 2000.

38. Searle (1995) defines social facts as those facts produced by virtue of relevant
actors agreeing that they exist. They rest on "collective intentionality." See
also Ruggie 1998.

39. For example, Wendt 1987, 1994; Dessler 1989; Kratochwil 1989; Katzenstein
1996.

40. Ruggie 1998:94, quoting Polkinghorne 1988.

41. See also Bernstein et al. 2000.

42. See Elster (1983:20:63–64) on the need for such a feedback mechanism in
functional theories. For example, "In biology the theory of natural selection
creates a presumption that whatever benefits reproductive capacity can also
be explained by these benefits." The social sciences lack a generalizable
mechanism of this sort, although a feedback mechanism may exist in particular cases.

CHAPTER 2

1. Conversely, deeply institutionalized norms, unless violated, may not be frequently articulated—they exhibit a taken-for-granted quality that can make
quantifying institutionalization difficult.

2. Unless otherwise noted, preparatory documents are found in U.S. Department of State 1972.

3. The exclusion of the German Democratic Republic, but inclusion of the
Federal Republic of Germany resulted from political maneuvering on the
part of the West. Neither Germany had UN membership, but the FRG was a
member of the International Atomic Energy Agency and a UN resolution in
1969 allowed its members to participate. UNCHE secretary-general Maurice
Strong (2000:120–121) however had fought earlier to appoint a leading Soviet scientist, Vladimir Kunin, to the secretariat staff, and a second Soviet expert joined even after the boycott. In addition, Strong personally met with
the Soviet ambassador in Stockholm every day of the conference to keep
Moscow informed of the proceedings.

4. Munn 1992 and R.E. Munn, author's interview. In chapter 4 I further discuss
Odén's role in developing the science that led to international responses to
acid rain.

5. The request that financial commitments from the developed world be in addition to existing development monies—often termed "additionality"—is
perhaps the one issue continually requested by developing countries that is
consistently rejected by some developed countries, particularly the United
States (although some inroads occurred at UNCED). It is for this reason that
additionality does not appear in the list of norms below. See Jordan 1994a.

6. Numerous sources detail NIEO demands and history. For example, Rothstein 1979 and Krasner 1985.

7. I use gender-inclusive language even when discourse at the time used gendered language such as "mankind." However, in quotations or in statements of principles such as the "Common Heritage of Mankind," I will use the original language on first reference for accuracy.

8. *Founex Report* 1972:32. As an illustration of how much has changed in the South, note that worldwide condemnation met a similar proposal made by World Bank Chief Economist Lawrence Summers in a 1991 internal bank memo (subsequently leaked). See Rich 1994:246–249.

9. The Agesta Group AB Sweden 1982. The report backs up its findings with a detailed breakdown of the implementation record of the 109 proposals in the action plan.

10. Strong 2000:118–119. See also Herter, Jr. and Binder 1993:2. For details of Strong's background, see his autobiography (Strong 2000) generally.

11. Rowland (1973:37) states that Strong's first decision was to dispatch Chester Ronning, a seasoned Canadian diplomatic trouble-shooter and China expert, to Beijing to meet with Zhou to convince the Chinese to attend. Rowland suggests that Strong's integrity played a large role in Zhou's acceptance. Rowland also argues that it might have helped that Strong could claim a distant relationship to the late Anna Louisa Strong, a left-wing American journalist sympathetic to the Chinese revolution.

12. However, the principle of advance notification, although in the draft declaration as principle 20, did not appear in the final document of June 16. (Brazil insisted it be put over to the General Assembly, for essentially short-term political reasons, as it was then embroiled in a dispute with Argentina to which the principle would apply.) Some developing countries also feared the principle could be abused by developed states to impede development projects. The Canadian delegation in its speech to the plenary argued that principle 20 still reflected a duty under existing customary international law. It did appear in watered down form in UNGA res. 2995 (XXVIII). It also appeared in various forms in later environmental agreements and the Rio Declaration entrenched it formally in principles 18 and 19. See Rowland 1973:99, 135–136; Grubb et al. 1993:89; Sands et al. 1994:8.

13. Sohn 1973:443–444. Sohn presents a paragraph-by-paragraph analysis of the Declaration that includes the evolution of wording in negotiations. The following discussion draws heavily on his summary.

14. United Nations 1972b.

15. The importance of the case to environmental law is cited in a wide range of publications. For example, see Sands 1994:xxxi.

16. On the one hand "rational" implies the use of instruments such as environmental impact assessments to set guidelines for development or to define "optimal pollution levels" (Colby 1990, 16–17). On the other hand, the OECD (1971) equated rational management explicitly with management in "accordance with basic economic principles." In other words, rational management concerns "how to internalize environmental effects in economic mechanisms so as to ensure a rational allocation of costs." Given the usage by conservationists in the 1970s and 1980s, I would argue that rational planning at Stockholm fits with the first view above.

17. The Agesta Group AB Sweden 1982:3–4. See also Colby (1990) and Sagasti and Colby (1993), who characterize the period following Stockholm as dominated by an environmental protection management "paradigm," defined as a reliance on legal regulations aimed to make short-term economic trade-offs to protect the health of people and a few species, and the separate treatment of environment and economics.

18. OECD 1975; Pearson 1994; author's interview, Jim MacNeill, a Canadian delegate and organizer of the OECD environment committee in 1970, and later director of the environment at OECD (1978–1984).

19. Turner (1995) argues that whereas environmental economists merely extend neoclassical economic theories to nonmarket phenomena, such as commons resources or public goods, ecological economists attempt to combine ecology and economics. For example, they attempt valuations of ecosystems and evaluate replacement costs—the cost to substitute artificial for natural processes that sustain a healthy environment.

20. Sands et al. 1994:xxxiv; Maurice Strong, author's interview; Pearson 1994:563; and personal observations of developing country speeches at the First Conference of the Parties for the Framework Convention on Climate Change, Ministerial Segment, April 4–7, 1995, Berlin.

21. Tolba officially took over December 1976, but unofficially began to run UNEP immediately after Strong stepped down in 1975 to head the newly formed Canadian Crown Corporation Petro-Canada.

22. A modified version of the speech is reproduced in Strong 1975.

23. The speech to the Second International Conference on Environmental Future, Reykjavik, Iceland, is reproduced in Strong 1977.

24. Strong 1977:170. Note, one difference between Strong and his successor is the former's spiritual vision of a world where intellectual, moral and cultural pursuits slowly take over from material pursuits in human development. Tolba left such inspirational speculations to others, instead focusing on education and implementation of UNEP's view of sustainable development.

25. See McCormick 1989:162–170 for a history of the drafting.

26. For example, sections 10.4.d, 13.4 and 13.5 on public participation in development planning and 14.10 and 14.11 on traditional knowledge in rural development.

27. UNGA resolution 38/161, para 8 a) and b) reproduced in YUN 1983:772.

28. YUN 1987:661–679. For a summary of UNEP's report see Dabholkar 1989.

29. Author's interview with a source who had high-level contact with WCED and UNEP.

30. On these competing environmental ideologies, see O'Riordan 1995a.

CHAPTER 3

1. Williamson 1993:1329. This should not imply that it originated solely in Washington. Williamson in fact cites the Latin American experience and intellectual trends there in response to the debt crisis as one of its more important sources, and notes it is part of a wider intellectual trend in development thinking.

2. Williamson 1990. See also Krugman 1995. Others present a slightly different mix. For example, Sachs' (1995) list includes open international trade, currency convertibility, private ownership, corporate ownership as the dominant organizational form, openness to foreign investment and membership in key international economic institutions. See also Pauly (1997:122) who emphasizes more the IMF's goal of opening financial markets. Notably, Williamson eschews the term neoliberalism or neoconservatism to describe the consensus because these terms include other policies that lack the same consensus. I thus use the term liberal economics as shorthand for a general trend away from state intervention and toward deregulation of markets and investment, privatization, liberalization of trade, and use of markets as a source of resource allocation.

3. Jordan 1994b. See also Fairman (1994) for evaluations of GEF's early performance.

4. See World Bank 1990 and later reports in the same series.

5. The World Bank (1992b:8) argued "there is no difference between the goals of development policy and appropriate environmental protection. Both must be designed to improve welfare." And, it sums up the rationale for its narrower definition as follows: "Basing developmental and environmental policies on a comparison of benefits and costs and on careful macroeconomic analysis will strengthen environmental protection and lead to rising and sustainable levels of welfare."

6. MacNeill, Winsemius, and Yakushiji 1991. Although it cannot be considered official Trilateral Commission policy, it was the Commission's main public response to Brundtland.

7. For a brief history of OECD activities, see OECD 1994a:11–25.

8. OECD Council (C(90)177/final) reprinted in OECD 1994a:11.

9. Dubiously, because the anarchical nature of the international system also means that market instruments would require global regulatory bodies with a high degree of legitimacy, surveillance ability, and political consensus. As chapter 5 will argue, such schemes do reflect a direction more consistent with existing international social structures, but that does not mean they are more practical, easier to set up, or more effective.

10. The United States, Japan, Great Britain, France, Germany, Canada, and Italy.

11. *Project 88* 1988; *Project 88—Phase II* 1991. See also *The Economist* 1988, 1991.

12. For an overview of trends in U.S. environmental policy up to Clinton's first term, see Vig and Kraft 1994. For a brief summary of the U.S. experience with economic instruments up to that period, see OECD 1994a, 1994b:295–298; Ingham 1994.

13. OECD 1994b:10–12. The authors point out the common property (as opposed to open access) as well as private property regimes may fulfill all four criteria. However, they favor private property since common property regimes, they argue, have a tendency to break down.

14. For succinct summaries of the agreements and negotiations see Grubb et al 1993. For full texts and preparatory documents see IDRC 1993. All references to United Nations', and other official, documents are from this source unless otherwise referenced.

15. In 1992, about 115 countries had environmental ministries or agencies compared with 11 in 1972. Imber 1994; Rogers 1993.

16. Dunlap, Gallup, Jr. and Gallup 1993. The authors admit that poorer, less economically developed nations, especially in Africa, are underrepresented, although the survey was the most comprehensive of its kind. The following is a list of countries surveyed organized by region: North America—Canada, United States; Latin America—Brazil, Chile, Mexico, Uruguay; East Asia—Japan, South Korea, Philippines; Other Asia—India, Turkey; Eastern Europe—Hungary, Poland, Russia; Scandinavia—Denmark, Finland, Norway; Other Europe—Germany, Great Britain, Ireland, Netherlands, Portugal, Switzerland; Africa—Nigeria. Less comprehensive surveys have been conducted. For example, Weale 1992, 25, notes that a Harris polling organization survey conducted in 1988–89 in 15 countries in all parts of the world found that leaders and publics in all but one country (Saudi Arabia) thought that the environment had become worse in the previous decade.

17. Parson, Haas and Levy (1992) estimate that about one-third of the approximately 1,400 NGOs accredited at the conference (as opposed to the parallel nongovernmental Global Forum) were from the developing world. See also Doherty 1994. Estimates range widely on the actual number of environmental NGOs in existence in the South.

18. For a summary of these and other pressures in the lead-up to UNCED see Brenton 1994:125–162.

19. A large number of books, articles, and speeches since about 1989 have supported the broadening of the security concept and particularly the notion of environmental security and/or the link between national security and the environment. See, for example, Woodrow Wilson Center 1996 and its subsequent publications.

20. For example, Rogers 1993; Chatterjee and Finger 1994; Grubb et al. 1993; Spector et al. 1994; Imber 1994; Campiglio et al. 1994; and Colorado Journal of International Environmental Law and Policy 1993. The best succinct summary is Haas, Levy and Parson 1992. Daily coverage of negotiations during preparations and the summit can be found in the *Earth Summit Bulletin* (which, following UNCED, became the *Earth Negotiations Bulletin*).

21. This includes not just environmental groups, but also industry groups, scientific organizations, and so on.

22. Some discrepancy exists in various reliable sources on the exact number of heads of states who attended (IDRC 1993 lists 104 speeches by heads of state) and NGOs at the Global Forum. The estimates are the most commonly cited.

23. The full title is the "Non-Legally Binding Authoritative Statement of Principles for a Global Consensus on the Management, Conservation and Sustainable Development of All Types of Forests."

24. For a detailed comparative analysis of attempts to institutionalize these two norms see Busumtwi-Sam and Bernstein 1997.

25. Based on a five-point definition in Rana 1994. See also Payoyo 1997; Schmidt 1989; Herber 1991.

26. However, the United States refused to sign the latter over the inclusion of CHP and controversy over a proposed management organization. Without U.S. support, the treaty has little impact. Rana 1994:247.

27. The search was conducted via a CD-ROM (IDRC 1993) containing all official UNCED and preparatory documents.

28. United Nations 1994. The "Agreement Relating to the Implementation of Part XI of the United Nations Convention on the Law of the Sea of 10 December 1982, with Annex, adopted at New York, July 28, 1994" (UNGA A/RES/48/263), passed by a vote of 121–0 with 7 abstentions, contains legally binding changes to Part XI and is to be applied and interpreted together with the Convention as a single instrument. See chapter 5 for specifics of some of the changes along liberal lines. The LOS Convention, Agreement, and related legal material and commentary, has been posted on the world wide web at http://www.clark.net/pub/diplonet/los.html. For other commentary, see AJIL 1994.

29. WCED 1987:67–91; Sands 1994:xxiv; Commission of the European Communities 1993:104–105; OECD 1994a, 1994b.

30. Smets 1994. See also Sands 1994:xxxiv.

31. Its legal status is subject to debate, but the institutionalization of its wider meanings of redistribution and shared authority is weaker than ever. Payoyo 1997:452–459.

32. I have glossed over technical debates on implementing PPP in trade agreements (see Pearson 1994). My point is not that all economic activities now incorporate PPP, but that as an environmental norm, its primary meaning constitutes the legitimate basis of linking trade, economic activity and environmental concern.

33. Sjöstedt et al. 1994:18. Hajost (1994) notes that the United States did not appoint a high-level point person for the negotiations and U.S. agencies, including the Environmental Protection Agency, did not make it a priority until very late in the negotiation process. This is in stark contrast to the leadership role played at Stockholm.

34. Chasek (1994b), who also gives a summary of the process, organization of issues, negotiations, and results achieved at the PrepComs. The discussion of PrepCom IV below draws primarily on her account.

35. See South Centre 1991, which served as the basis of common negotiating strategy for the South. See Porter and Brown (1996:117) for a summary. On the South's negotiating stance more generally, see Mensah 1994.

36. For a detailed discussion of the Precautionary Principle and its history see O'Riordan and Cameron 1994. See also Bodansky 1991.

37. See, for example, Costanza and Cornwell 1992. See also OECD (1994a:43, 149), which invokes the principle (along with PPP) to legitimize increased use of market instruments.

38. Ricupero later became UNCTAD's fifth Secretary-General (in 1995). As Brazil's finance minister, he supervised the launching of the Brazilian economic stabilization program in 1994.

39. See Ricupero 1993 for his account of the negotiation of the financial chapter of Agenda 21.

40. For example, the Montreal Protocol allows a ten-year grace period for developing countries.
41. At the First Conference of the Parties for the Framework Convention on Climate Change in Berlin (1995), the G-77 formally split when India, at the behest of small island states, led the majority of developing countries to push for a strong protocol to limit emission of greenhouse gases, over the objections of a number of oil producing states.
42. On the negotiation of the Declaration see Kovar 1993; Imber 1994; Grubb et al. 1993:85–95; Porras 1994; Campiglio, et al. 1994. Many of the same issues arose in negotiations over climate change, which have been documented in a large number of books and articles. See, for example, Mintzer and Leonard 1994; Rowlands 1995; Paterson 1996.
43. The quotation is from Principle 12 of the Declaration.
44. Indeed, international lawyers did not play a prominent role in the negotiations over the proposed Earth Charter/Rio Declaration. Significantly, there was no suggestion that the International Law Commission play a prominent role in the Earth Summit preparations, an indication that most states felt the creation of new international law was mainly a political process best left to intergovernmental negotiations. The sense among international lawyers at that time was that the law commission is trusted to refine international law, not to develop new law, especially in a process like the Earth Summit where political, economic, and scientific considerations take precedence. Shibata 1994.
45. In 1994 parties to the 1989 Basel Convention on the Transboundary Movement of Hazardous Waste and its Disposal agreed to a ban on the waste trade with less-industrialized countries, although at the time of writing the ban lacked the necessary ratifications to enter into force. The ban is an exception to the general thrust of Principles 12, 13, and 14, which support trade liberalization as a way to improve environmental quality. On the ban see Krueger 1999; Clapp 1994.
46. The text of the protocol is available from the UNEP website at http://www.unep.ch/basel/COP5/docs/prot-e.pdf. For a discussion and summary, see IISD 1999.
47. The Ministerial Declaration is available at http://www.unep.ch/basel/COP5/ministerfinal.htm.
48. The United States released "interpretive statements" (one of only a handful of states to do so) on some sections of the Rio Declaration and Agenda 21. They included the U.S.'s long-standing opposition to a "right" to development on the grounds that a "right" might override other rights, such as human rights. According to the statement, the United States does not oppose principle 3 understood as the promotion of development "in a way that the development and environmental needs of present and future generations are taken into account." The statement on principle 3 is reproduced in Thompson 1993:90 fn. 1.
49. However, the principle limits its application to national legislation and projects that are subject to "a competent national authority." However, Principle 19—advance notification—implies states ought to notify others of results of an assessment that might produce effects beyond state borders. Hence, the

Declaration implies transnational application of the principle. See Kiss 1994:60.

50. For a summary see UN Commission on Sustainable Development 1997. The Biosafety Protocol is discussed further in the concluding chapter.

51. Other initiatives included the Helsinki Process on protecting forests in Europe, the Montreal Process on creating criteria and indicators (C&I) for the conservation and sustainable development of temperate and Boreal Forests (the Helsinki process included a C&I process as well), negotiations toward a successor to the 1983 International Tropical Timber Agreement (which led to a new agreement in 1994), and initiatives on labeling and certification schemes (not to be confused with C&I). The most important certification scheme, under the Forest Stewardship Council, involves market players and NGOs who want to set criteria to identify products produced from "well-managed" forests. A large number of expert and governmental workshops also convened on various aspects of the forestry issue. See Humphreys 1996 for a summary.

52. This section draws primarily from Humphreys 1996 and summaries of the IPF process from various issues of the *Earth Negotiations Bulletin*, volume 13 (on forests) unless otherwise noted.

53. *New York Times* 1997:7. Those who expressed support for a convention included Canada, EU, Indonesia, Malaysia, the Philippines, Papua New Guinea, China, Costa Rica, Poland, the Forest Alliance of British Columbia and the Canadian Pulp and Paper Association.

54. *New York Times* 1997:7; and personal correspondence with a Greenpeace representative on forests, San Francisco.

55. See Paragraph 28 on production and consumption and Section C on implementation in UNGA 1997.

56. The May 1995 Report on Trade and Environment to the OECD Council at the Ministerial Level (and accompanying studies) and the November 1996 Report of the Committee on Trade and Environment to the WTO Ministerial Conference in Singapore are summarized in Reiterer 1997. The quotation comes from the OECD report, quoted in Reiterer 1997:72.

57. Without formal agreement on PPMs, trade law on this issue is evolving through the (so far) few trade-environment disputes that touch on the issue, including "tuna-dolphin" and "shrimp-turtle." See, for example, Wynter (1999) for an argument that the WTO Appellate Body's decision on the "shrimp-turtle" dispute between the United States and India, Malaysia, Pakistan and Thailand, over the danger to protected sea turtles by shrimp trawling, left the door open for the use of PPM-based trade measures, if properly applied, under GATT Article XX, even though it ruled against a U.S. restriction on imports of shrimp caught with nets that are not equipped with "turtle excluder devices." The ruling found the U.S. acted in an unjustifiably discriminatory manner, not that PPMs were impermissible.

58. At meetings of the CTE during 2000, many delegates asked that the committee work on clarifying coverage of eco-labels under the Technical Barriers to Trade Agreement, and some expressed concern they could become an unnecessary barrier to trade. At the same time, delegates supported their use as an effective tool to promote environmental policies. One member state ar-

gued that eco-labels that included non-product-related production and processing methods were not consistent with WTO rules. There is general agreement within the CTE that eco-labeling schemes (whether mandatory or voluntary) should be developed in a transparent, nondiscriminatory (e.g., consistent with rules of national treatment), and least trade restrictive manner to achieve the policy objective. WTO 2000. Pressure may build to only allow mandatory schemes, such as those agreed to under multilateral environmental agreements, since voluntary schemes run a greater risk of being considered barriers to trade under WTO rules. However, agreement on mandatory schemes is obviously much more difficult.

59. Although many forest companies are beginning to show an interest in certification, some skepticism remains as to whether the process can be adequately reconciled with sustainability goals at the national or international level, whether certification might be used as a non-tariff barrier, and whether consumer demand will create sufficient incentive for companies to take part in such schemes. See Gale and Burda 1997; Kiekens 1997; Hansen 1998; and Bernstein and Cashore 2000b.

60. The text of the Protocol and Convention can be downloaded from http://www.unfccc.de.

61. For a review of trading under the Montreal Protocol see OECD 1998:25–33.

62. "Developed" here means Annex B countries of the Kyoto Protocol—those states that face quantified reduction and limitation commitments under the Protocol. Annex B includes all OECD countries except recent members Mexico and South Korea, and Turkey because it had not ratified the Framework Convention on Climate Change when Kyoto was signed. Annex B also includes economies in transition. Developing countries refer to non-Annex B states.

63. See IISD 1997b for a summary of negotiations. See also Grubb, Vrolijk, and Brack 1999 for a detailed analysis of the Protocol and its implications.

64. Negotiations at the Hague broke down primarily over the issue of carbon "sinks" or sequestration of carbon in forests or by other land-use or agricultural changes (under articles 3.3 [afforestation, deforestation, and reforestation] and 3.4. [additional activities that sequester carbon]. The Protocol allows the use of sinks, but they remain controversial, particularly owing to questions of accounting and verifiability, and whether the inclusion of sinks threatens the Protocol's environmental integrity. Specifically, states could not agree on whether to allow activities under article 3.4 in the first commitment period (2008–2112—the dates by which developed states agreed to meet their first reduction or limitation commitments under the Protocol) or whether or under what conditions sinks could be included in the CDM. The main protagonists were the Umbrella Group, which includes the United States, Canada, Australia, New Zealand, Japan, Norway, Russia, and Ukraine, who fought for wide latitude on the inclusion of sinks, and the EU, which opposed this position. A variety of technical issues related to sinks and the Kyoto mechanisms also remained unresolved, as did the shape of the compliance mechanism and questions regarding financing for developing countries. See IISD 2000e for a summary of COP-6.

65. As this book was going to press, U.S. President George W. Bush announced his country would withdraw from Kyoto, despite its 1998 signature. Since entry into force requires ratifications by 55 parties that account for 55 percent of developed country emissions, U.S. ratification (its emissions account for 36.1 percent) is important, but not mandatory. Indeed, most developed states have indicated they remain committed to action consistent with the Protocol. While the future of Kyoto is uncertain, some developing countries, led by Argentina (IISD 1998), have already pledged to undertake voluntary commitments, one condition for U.S. ratification. Since Bush announced no alternative proposals, any changes are likely to be toward even stronger allowances for market mechanisms and greater flexibility on targets, further reinforcing liberal environmentalism.

66. For example, as mentioned earlier, Shell and BP pledged to cut emissions (ENS 1998), and later BP Amoco (following a merger) and Royal/Dutch Shell launched internal carbon emission trading markets to help them reach their pledge (*Financial Times* 2000). In addition, organizations ranging from UNCTAD to the UN Industrial Development Organization have conducted research or workshops on the various mechanisms and the roles they can play in implementation. For example, on UNCTAD's activities, see its Greenhouse Gas Emissions Trading website at http://www.unctad.org/en/subsites/etrade/index.

67. Chatterjee and Finger 1994; Sachs 1993; Rogers 1993; Maurice Strong, author's interview.

68. The BCSD actually lobbied both positions in some ways. For example, it sought the removal of references to regulation of multinational corporations from Agenda 21, but also promoted the idea of voluntary self-regulation for industry.

CHAPTER 4

1. These two questions, for example, are ably addressed elsewhere, including, recently, by Andresen et al. (2000) on the former, and Meyer et al. (1997) on the latter. Neither, however, focuses on the equally important question of the normative content or framing of policies produced, that is, why global environmental problems are being managed or addressed in the ways they are.

2. Epistemic communities, for example, are especially noted for their progressive influence on "learning" within international organizations and domestic bureaucracies. See Haas and Haas 1995.

3. See Goldstein and Keohane 1993b:11; Yee 1996; Litfin 1994; Haas 1992b on its influence on the ideas literature in international relations and comparative politics.

4. On epistemes and the earliest formulation of the concept of epistemic communities, see Ruggie 1975, who draws especially on Foucault. On the broader potential agenda of the epistemic communities literature, see Adler and Haas 1992.

5. For example, the inclusion of scientific advisory panels for a number of treaties has affected decision making on technical matters, although to varying degrees. See Andresen et al. 2000; and Deutz 1997.

6. Litfin (1994) and Susskind (1994) raise similar issues that suggest a more complicated path through which science affects international environmental policy.

7. For a detailed, but very critical, review of the new sociology of science, see Bunge 1991.

8. A large literature exists on many of these concepts. For brief summaries that touch on their transnational applications see Bennet (1991:224–225) on policy communities and networks; Sikkink (1995) and Keck and Sikkink (1998) on transnational issue or advocacy networks; Smith et al. (1997) on transnational social movements; and Sabatier and Jenkins-Smith (1993) on advocacy coalitions.

9. See Hagen 1992:64–65, 122–145. The classic text is Odum 1953. Ecosystem and systems ecology are not always considered the same, but are associated owing to Odum's influential work.

10. William Clark, author's interview. Note, Clark, who has led major transnational research programs on sustainable development, does not refer to himself as an ecologist, but was trained as one and works on international ecological issues.

11. For these trends in global change research see *Ambio* 1994.

12. This is not an exhaustive list, but simply examples of individuals who led in promoting global environmentalism. Among other things, Strong served as secretary-general of the Stockholm and Rio conferences and as a commissioner on the Brundtland Commission; MacNeill was secretary-general of the Brundtland Commission and director of the OECD environment committee 1978–84, Thacher served as deputy director of UNEP 1977–1983 and occupied a number of roles in U.S. environmental diplomacy since the early 1970's, and Tolba headed UNEP from about 1977–1994.

13. Adams 1990:33. On the IBP see Hagen 1992:164–188.

14. Kowalok 1993:13–14. The discussion of acid rain draws primarily on Kowalok; Munn 1992 and author's interview with Munn.

15. From there to regional cooperation on mitigating acid rain in Europe and North America is a larger, and more complicated story—the illustration is merely to suggest that producing political action often stems from factors unrelated to scientific consensus. However, evidence also suggests that major turning points in reaching interstate agreements on acid rain stemmed more from political factors independent of new scientific evidence, such as Germany's 1982 sudden about face owing to a public outcry and media coverage over damage to the Black Forest attributed to the problem. Increased political salience then contributed to increasingly coordinated transnational scientific efforts and institutional development as much as vice-versa, and policy prescriptions (from specific targets to the tradable permit system for sulfur dioxide emissions in the United States) often did not stem from scientific findings, but from political considerations and other policy norms. On

the larger story of the politics-science interaction on acid rain cooperation, see, for example, Wettestad 2000. For an analysis focusing on Canada-U.S. cooperation, see Munton 1998.

16. This meeting was part of the First General Assembly meeting of SCOPE, Aug. 29–Sept. 4, 1971. At the same meeting, scientists in SCOPE gave Strong his first scientific briefing on problems of the environment. ICSU 1971:15–17; R.E. Munn (author's interview), a founding organizer of SCOPE.

17. For a summary of the role of scientists in some specific recommendations, see Thacher 1973; Kellermann 1973.

18. *IUCN 1972 Yearbook*: 20, quoted in McCormick 1989:98.

19. SCOPE 1971. According to Ted Munn (author's interview), the actual funding proposal was written in one night after Tom Malone brought Munn and British scientist Gordon Goodman to UNEP headquarters and told them if they got a funding proposal on Maurice Strong's desk by eight o'clock the next morning, they would get their money.

20. Author's interview. For Sachs vision of ecodevelopment, see, for example Sachs 1977, 1984.

21. I discuss the failure of ecodevelopment further in chapter 5.

22. For example, see sections 10.4.d and 13.4 and 13.5 on public participation in development planning and 14.10 and 14.11 on traditional knowledge in rural development.

23. The United States especially dominated research on ozone depletion. Haas 1992d:193.

24. Unless otherwise noted, information on the scientific history of the ozone issue is drawn from Kowalok 1993; Litfin 1994; Brodeur 1986; Rowlands 1995a:43–64.

25. Interpretations vary on the policy effects of the report from the U.S. government sponsored Climatic Impact Assessment Program, which involved more than 1,000 scientists from 10 countries, and on the precise reasons for the cancellation. Rowlands (1995a:45–46) argues that the report was open to wide interpretations because its executive summary, on which policymakers relied, focused on the minimal threat from a small fleet of SSTs and the report made only oblique references to more severe consequences in scenarios found elsewhere in the report. See also Litfin 1994:62; Kowalok 1993:17.

26. Note, in support of a scientific basis for the Precautionary Principle, Lemons and Brown (1995:20–21) argue that precaution stems from a "holistic" approach to science as opposed to positivist, predictive science. The former—because it is sensitive to complexity, uncertainty and interaction with other ecosystems and human activity—focuses on avoiding type II statistical errors (false negatives). Traditional predictive science focuses on minimizing type I statistical errors (false positives). However, most laboratory scientists who take a formal view of science argue the contrary, that such a position is not "scientific." There is certainly nothing resembling consensus on this point within relevant scientific communities.

27. Thomas quoted in Litfin 1994:104, see also 72–73. Brodeur (1986:78) notes that the previous administrator, Anne Gorsuch, took a very different view.

She called the science "highly controversial" in her Senate confirmation hearings. She also said that there was a "need for additional scientific data before the international community would be willing to accept it as a basis for additional government action."

28. Litfin 1994:99. On NGO support of precaution on other issues see Princen and Finger 1994:20, fn. 81.

29. Robert Watson, unpublished interview with Peter Berry (University of Toronto), January 1996, Washington, D.C.

30. Robert Watson, unpublished interview with Peter Berry (University of Toronto), January 1996, Washington, D.C.

31. Robert Watson, unpublished interview with Peter Berry (University of Toronto), January 1996, Washington, D.C.

32. Clark and Munn 1986. Munn (1987) lays out the intellectual framework for the environment program at IIASA.

33. William Clark, author's interview. On the use of these concepts at IIASA see Clark and Munn 1986; Munn 1987.

34. Some discrepancy exists in the actual number of reports submitted. Haas, Levy and Parson (1992:11) count 130 reports; IDRC (1993) lists 164 national and 13 regional reports; and Grubb et al. (1993) count 172 by UNCED, a number that rose to 190 by the end of 1992.

35. Dooge et al. 1992. See also "Recommendations from Sigma Xi and ASCEND 21" 1992; Marton-Lefèvre 1994.

36. Interview of a participant in the ASCEND conference.

37. South Centre 1991; Porter and Brown 1994:122; Arizpe, Costanza, and Lutz 1992; Grubb et al. 1993:30–33; *The Economist* 1992.

38. Susskind 1994:62. On the lead-up to the treaty see McConnell 1996; Brenton 1994:197–206; Grubb et al. 1993:75–84; Boyle 1996.

39. Rowlands 1995a:85–87; Grove 1991:66–67, adapted from his article in *Nature* (May 3, 1990). For a history of climate change research, focusing especially on the United States, see Hecht and Tirpak 1995; Victor 1995.

40. For example, Boehmer-Christiansen 1994a:146 fn. 22 cites, in addition to Bolin, Paul Crutzen, Sir John Houghton and Professor Yuri Izreal, who have served variously as WMO executive members, IPCC Bureau members and chairs of IPCC Working Groups, among other posts. However, more than one scientist I interviewed also mentioned bureaucratic competition between IPCC and WCRP.

41. Boehmer-Christiansen 1994a. The combination of climate research and energy policy, especially in alternatives such as nuclear energy, is a key theme in Boehmer-Christiansen's analysis.

42. Note, while Munn identifies Clark's keynote address as the paper that galvanized scientists into action, Clark himself downplays his role, or that of any single paper, and attributes conference chair Jim Bruce's framing of the policy implications of climate change as the main catalyst. Bruce framed the issue from the perspective that crucial economic decisions of governments, and the energy and resource sectors, were being made under the assumption that climate would remain relatively unchanged, and that since science now suggested that assumption no longer held, what might be the implications of

that and what should government and industry do next? R.E. Munn, William Clark, and Jim Bruce, author's interviews.

43. WMO organized the 1979 conference to demonstrate success in its long-standing research program on weather forecasting, so it turned to climate change, at least in part, to show atmospheric research still deserved funding. Subsequently, GARP transformed into the World Climate Research Programme, with support from interested states including the United States, Soviet Union, United Kingdom, and Canada. Author's interview, Gordon McBean, who has held executive positions at WMO, ICSU, and WCRP.

44. Boehmer-Christiansen (1994a:156) uses the term "science managers;" Litfin (1994) uses "knowledge brokers."

45. See Boehmer-Christiansen 1994a:156, who lists key individuals comprising the network who remained major players in scientific research up to and including the IPCC process. Names that stand out include Bolin, Tom Malone from ICSU and WMO, Gordon Goodman from the Stockholm Institute, and Tolba.

46. That estimate, however, is increasingly seen as conservative. The draft IPCC Working Group 1 Third Assessment Report released in January 2001 projects global mean surface temperatures to rise by 1.4–5.8 degrees C (2.5–10.4 degrees F) from 1990 levels by the end of this century based on GHG emission trends, a substantial increase over 1995 projections. (The main difference is based on revised estimates of the influence of sulphates from industry and power plants). It also takes a firmer position that anthropogenic (human-induced) emissions have contributed to oberved warming over the last 50 years (IPCC 2001).

47. Agrawala (1999:160–162), who also presents a detailed history and influence of the AGGG; interview with James Bruce, deputy secretary-general of WMO when AGGG was created.

48. The following discussion of the Toronto conference draws primarily from Agrawala (1999) and Paterson (1996:33–34).

49. Haas, Levy, and Parson 1992:10, make a similar point. For one of the only detailed studies of the role of such expert bodies on adequacy of commitments within treaties, see Deutz 1997.

50. James Bruce, co-chair of working group III of the IPCC second assessment report, author's interview. The list of authors for working group III includes some natural scientists, but primarily prominent economists including David Pearce of the United Kingdom and nobel laureate Kenneth Arrow of the United States.

51. Note, Elizabeth Dowdeswell, who later replaced Mustafa Tolba to head UNEP, is credited with the idea to revamp working group III. Dowdeswell's background is primarily as an administrator/bureaucrat, unlike Tolba who gained a reputation first as a scientist.

52. Gordon McBean, author's interview. When WMO set up IPCC, Japan sent representatives from its Ministry of International Trade and Industry, not hard scientists, and expressed a desire to get involved on response strategies because it saw opportunities as a leader in solar cell and other alternative energy technologies.

53. When controversy erupted over these figures, they were removed from the summary report, although they remained in the actual chapters. For brief notes on the controversy, see Skodvin 2000:162–164; "Heating up the Climate Change Debate" 1996; Wysham 1994.
54. WMO/UNEP 1992.
55. For example, Bolin 1994. Also, Sherwood Rowland, in Brodeur (1986:80–81) lists a number of reasons why scientists are not more active.
56. For example, ICSU and SCOPE generally failed to incorporate views of social scientists.
57. Susskind (1994:62–81) summarizes a number of these problems.

CHAPTER 5

1. I use the term policy entrepreneur generically to refer simply to individuals who creatively attempt to change, reframe, or promote policies or norms. While not inconsistent with labels such as norm entrepreneur or moral entrepreneur used by others (Nadelmann 1990; Finnemore and Sikkink 1998), I use a broader label here to avoid making a causal claim about the source of authority or reason for success of such individuals or groups independent of the broader theory proposed here.
2. Pigou 1920:23–30. See also Weale 1992; Hahn and Stavins 1992.
3. For a brief list of prominent studies see Thompson 1972. Notably, one such study cited was commissioned by the Canadian government from Jim MacNeill (1971), who later became secretary-general of the Brundtland Commission.
4. Although Moravcsik (1997:517 fn. 6) insists his theory of preferences is not pre-social, it must be if he assumes fully exogenous preferences.
5. Pioneering works in the new institutionalism include March and Olsen (1984, 1989). See also their foray into International Relations scholarship (1998).
6. This explanation should not be confused with attempts to superimpose a model of evolution from biology onto social phenomena. Efforts to develop social scientific concepts *directly* analogous to evolutionary mechanisms and derived from specific theories of evolution in biology have been employed to explain, for example, transformations in the international system or institutions, or conditions for optimum performance in the international political economy. See Modelski and Poznanski 1996 and other contributions to the September 1996 special issue of *International Studies Quarterly*. Here the analogy is strictly limited to the idea of fitness as a contingent phenomenon.
7. Florini also emphasizes legitimacy, but discusses it as indicative of norm "prominence," one of the factors that influences whether a norm will be accepted (1996:374–375). She uses "prominence"—a direct analogy to gene prominence as a factor that accounts for reproductive success in evolutionary theory—to encompass a number of other factors that affect the promotion of a norm. Thus the analogy, although creative, is too forced to focus attention directly on social processes. Spruyt (1994) also acknowledges a role for legitimacy, but gives greater emphasis to efficiency in his theory of the evolution of institutional forms in international politics. The main driver of

selection in his theory is the relative efficiency of political units to achieve given goals—i.e., reducing transaction costs, raising revenues, preventing free-riding—over their rivals in a given environment. Mutual acceptance of the legitimacy of some institutional forms over others plays a secondary role, although it is one of three factors he identifies as important in the selection process. See especially pp. 158–169, and 201 fn. 47.

8.   My emphasis on this interaction also differentiates my approach from Emanuel Adler's theory of "cognitive evolution" (1991). Adler focuses primarily on innovation and learning within states, and secondarily on the subsequent diffusion and selection of new interpretations of the social world domestically and internationally. My approach focuses more directly on the interaction of ideas (whatever their source) and social structure.

9.   I am only focusing on new ideas about global problems as a source of change. Social structure may also change owing to the emergence of new identities, such as from revolutionary states, changes in technology, wars, and so on. However, in all cases, I would argue, social structure reflects human beings' construction of meanings around such events and social forces, not simply from those forces or technological changes themselves. Technological change no more determines social structure than vice-versa, but the two interact. Spruyt (1994:21) makes a similar argument, as have a large number of social theorists, at least since Max Weber.

10.   Florini 1996:364–365; Franck 1990:16, 38. Such a view of social structure is well supported within the constructivist literature. See Wendt 1992, 1994; Wendt and Duvall 1989; Dessler 1989; Kratochwil 1989; Busumtwi-Sam and Bernstein 1997; among others.

11.   For example, Franck 1990; Frost 1996; Kocs 1994; Ruggie 1998; Reus-Smit 1997; Wendt and Duvall 1989; and Kowert and Legro 1996 all conceive of levels of norms. However, the specific formulation below of a three-tiered hierarchical structure is an innovation of Busumtwi-Sam and Bernstein 1997, and the following discussion is based on the model of social structure developed there.

12.   This does not mean everything associated with the state is socially constructed, which is a claim virtually no one makes. Buildings, military weapons, flags, and documents of a state and the rocks, trees, water, and territory on which they sit are real. Rocks and trees are ontologically objective. Borders, however, are subjective. They define the boundary of the territorial state, and would not exist without the institution of sovereignty, or would at least mean something quite different. The meanings and uses of objects associated with the state are not self-evident as brute facts—the institutions of statehood are historically and socially constructed. The objects listed above are not inexorably linked to an entity called the state or a particular understanding or manifestation of sovereignty or political organization more broadly (although the symbolic content of these objects may be). States and international institutions are constructed in the sense that understandings of the state and sovereignty depend on shared meanings, they are socially constructed "all the way down," since even the most basic principles underlying sovereignty are not timeless, but came about through human interaction and understanding.

13. Ruggie (1986:144–145) argues property rights—the basis upon which states own property (territory) "differentiate among units in terms of possession of self and exclusion of others." Second, they designate a form of society—like private property rights, sovereignty establishes a society of "possessive individualists" arranged to reproduce this mode of differentiation and facilitate orderly exchange among separate parts.

14. Dessler 1989 among others. Krasner (1988:81), also uses an evolutionary analogy to argue that an explanation of how institutions begin needs to take account of the "genetic stock" of extant institutional structures, "not just external factors."

15. The practices of sovereign states are primary because the contemporary international system empowers states as primary units.

16. As Adler notes, quoting Popper, "in contrast to the past which is closed, as it were, the future is still open to influence, it is not yet completely determined" (1997:350 fn. 18).

17. I have not proposed a theory to explain the full relationship between power, external factors such as technological change, and social structural change and continuity, which is why the explanation relies on a historical snapshot of social structure at particular times to explain the entrance of new norms. As Finnemore (1996:15) notes, however, "There is no reason why the structure in a structural argument must be material and economic. Structures of shared knowledge and intersubjective understandings may also shape and motivate actors," and can be useful starting points for analysis.

18. Maurice Strong, author's interview. Subsequent quotations in this paragraph and the next are also from this interview. See chapter two for a list of participants.

19. Maurice Strong, author's interview. Sachs, however, is a critic of neoliberal or "scientific" economics and classifies his work as turning back toward a political economics or perhaps to a new "anthropological" economics. I. Sachs 1984:vii–ix. Also, I Sachs 1977.

20. Riddell (1981:8–14) lists 11 "macro principles" of ecodevelopment to guide such policies: (1) establish an ideological commitment (to ecodevelopment); (2) increase social equity; (3) attain international parity; (4) alleviate poverty-hunger; (5) eradicate disease and misery; (6) reduce arms; (7) move closer to self-sufficiency; (8) clean up urban squalor; (9) balance human numbers with resources; (10) conserve resources; (11) protect the environment.

21. McCormick (1989:162–170) details the complex compromises between various environment and development communities in the strategy's drafting.

22. For example, OECD 1994a, 1995, and 1998. The latter is one of many studies under the auspices of the Annex 1 Expert Group on the Framework Convention on Climate Change, whose analysis has been an important source of learning and influence in FCCC negotiations, especially on issues such as emission trading.

23. Jim MacNeill, author's interview. Hajer (1995:97–99) also credits the conference and OECD activities from 1979–84 as one of the most important sources and disseminators of these ideas.

24. MacNeill 1984 and author's interview. Strong (author's interview) did not see the choice between the possible agendas in as stark terms as MacNeill did, but rather as a continuation of the work started at Stockholm to link environment and development. He called the decision "tactical," and said that he did not recall the commissioners having a lengthy substantive discussion. As he tells it, "Jim and Gro Brundtland consulted with a few of us [and] felt that was the way we should orient our work . . . and it was decided that the new agenda was the better framework."

25. The argument parallels Finnemore's argument (1996:98–99, 114) crediting the World Bank's promotion of poverty reduction with giving legitimacy to this goal, which then influenced a shift in intellectual development circles toward a poverty orientation.

26. Weale (1992:23–32) and Hajer (1995) both focus on the "discourse" of ecological modernization.

27. Williamson 1990, 1993; Biersteker 1992; Rodrik 1994. See chapter 3 for some specific policy prescriptions attached to the "Washington Consensus," especially in IMF and World Bank programs. I am less concerned with specific policy advice—for example, whether pegged or fixed exchange rates should be preferred to floating rates—and more by the acceptance of broadly liberal economic norms in formerly recalcitrant countries.

28. For a historically grounded account of the IMF's promotion of these policies, see Pauly 1997. He argues that the Fund staff did not show particular intellectual creativity—indeed the policies themselves were little different than the League of Nations attempts to restructure debt in Central Europe following World War I—but rather used the Fund's surveillance function and financial resources in order to reduce bank debts in leading states while avoiding expulsion of debtors from international capital markets. The Fund provided incentives for its policies (see chapter 3) with the promise of restructured and reduced debt, and new financing in the form of private portfolio and direct investment.

29. See McNamara (1998) for a discussion of the intellectual and political roots of these economic norms in Europe toward a neoliberal consensus. According to McNamara, these policies placed price stability above all other goals as necessary for the success of the European Monetary System in the 1980s. She argues that these policies prevailed owing not to ideology, but to the failure of Keynesianism to respond to deteriorating economic conditions following the first oil shock in 1973, policy innovation along monetarist lines, and emulation, particularly of Germany, where restrictive monetarist and anti-inflationary policies appeared effective. On the developing world acceptance of these norms, see Biersteker (1992); Busumtwi-Sam (1995); and Rodrik (1994) for various explanations.

30. For G-7 summit statements during this period see Hajnal 1989, 1991. For a summary of international statements in support of market instruments see OECD 1994a:13.

31. For a summary of international initiatives, see for example, OECD 1994a; *Project 88—Round II* 1991:2–4. For Post-UNCED programs see, for example, OECD 1994b, 1995, 1998.

32. Descriptions of specific initiatives on climate change can be found at UNC-TAD's Greenhouse Gas Emissions Trading website, www.unctad.org/en/sub-sites/etrade/index.htm. For a general examination of UNCTAD's attempts since 1992 to incorporate environmental concerns into its research and programs, see Arda 1996.

33. UNGA 1994. The report was written under then Secretary-General Boutros Boutros-Ghali, but refers to comments made by de Cuellar in 1990.

34. For a detailed discussion see Busumtwi-Sam and Bernstein 1997. Among other changes, the 1994 agreement changed the role of the proposed management organization (the Authority) so that it now must respond to the "right" of states to mine mineral resources in the deep seabed, essentially putting in place an assured access rather than a common heritage regime. This change severely limits, perhaps even eliminates, the ability of the weakened Authority to control access. In addition, technology transfer provisions were changed so that transfers will now be guided by "conditions on the open market" and on "fair and reasonable commercial terms," and development of the Seabed shall proceed according to "sound commercial principles." See also Payoyo 1997:457.

35. Deudney does not argue that such a shift is occurring; rather, he sees in "green culture" some "of the major ingredients lacking in previous cosmopolitan alternatives to nationalism."

CHAPTER 6

1. Although I have not attempted it here, even apparently exogenous shocks might be fruitfully explored within a social structural framework since even major wars occur in the context of existing social relations.

2. Ruggie 1995; Wendt 1998. Admittedly, I have skirted over the issue of the relationship of the social world to the material world. I would argue, however, that investigating the content of social structure historically is sufficient methodologically for this approach, even if social structure ultimately interacts with material structure. This sort of chicken and egg problem is likely an unsolvable dilemma of social life. However, accurately identifying social structure serves as a useful shorthand for the manifestation of material interests, which, I have argued, are meaningful only in the context of social structure in any case.

3. See note 55, chapter 4.

4. James Bruce (author's interview) noted that economists in IPCC working group III for the 1995 report had a far greater consensus on core policy questions than the natural scientists involved.

5. One exception is Robert Keohane's review (1978) of the influential McCracken Report, although he looks more at the report than at the OECD's power *per se*. A few examples of systematic research on other organizations that might fall into this category include Gill (1990) on the Trilateral Commission and Pauly (1997) on the IMF, especially his focus on its surveillance function.

6. Finnemore 1993, 1996; Sikkink 1993; Keck and Sikkink 1997. See also Bernstein and Cashore 2000a.

7. See also Woods (1995:168) who focuses on the limitations of the "discovery" view of new ideas—that policies change when new good ideas are discovered.

8. Helleiner (1996:70) is referring specifically to works such as Haas, Keohane, and Levy (1993) and MacNeill, Winsemius, and Yakushiji (1991).

9. Hirschman 1989:349. Although Hirschman presents this rational response argument in the case of Keynesian economics, the Hall volume (1989a) as a whole finds it unconvincing. Hall (1989b), for example, argues that this "economist-centered" explanation is incomplete since economic theories and the economists that supported them waxed and waned in influence in comparison to other influences on policy. He argues for a more complete model of the policy process that looks at the interaction of ideas with their policy environment—administrative, political, and economic.

10. McNamara 1998. For example, she argues, in part, that the apparent success of monetarist policies in Germany in response to stagflation (simultaneous inflation and recession) in the 1970s, and the perceived failure of Keynesian policies, led other European countries to imitate Germany.

11. For example, Grubb (1998), among other concerns, questions the economic theories underpinning the Kyoto "flexible" market-based mechanisms even on efficiency grounds. He shows that "least cost" solutions initially do not necessarily lead to efficient outcomes in the longer term because, for example, the assumed incentive to innovate could be lost under such schemes.

12. IISD 2000d. Robert Watson, director of the World Bank's Environmentally and Socially Sustainable Development Program, also told the World Bank European Sustainable Development Forum that Bank President James Wolfensohn is placing greater emphasis on the promotion of sustainable development and that the identification of trade-offs is the main feature that distinguishes the 2002 strategy from the 1992 strategy.

13. Adding these criteria further complicates the already highly uncertain and politically charged science of modeling the costs and benefits of actions to mitigate climate change. Uncertainties range from energy prices and market conditions, to technological innovations. For example, models that allow scope for cost-effective improvements in energy efficiency might translate into net benefits (i.e., no-regret measures). Whereas such models face criticism for underestimating hidden costs of implementing new technologies, top-down models typically used in national forecasts tend to ignore such measures altogether, probably overestimating costs. See Grubb et al. 1999:163–165 and Appendix 2.

14. Anderson and Leal 1991. Note, however, that the supporters of "free market" environmentalism seem to misunderstand many of the compatibilities of current formulations of sustainable development with their position, thus they set up any environmental intervention as a target for attack. See Eckersley (1993) for a critique of free market environmentalism, even when it is less radical than the version endorsed by Anderson and Leal, as inappropriate as a blanket solution because it cannot balance economic efficiency with equally important goals of social justice and ecological sustainability.

15. See for example MacNeill, Winsemius, and Yakushiji (1991) where all these proposals are brought forward, yet political support has been found mainly

for action on proposals consistent with liberal/market norms. See also Weale (1992:157) and Goodland, Daly, and Serafy (1993) who argue for the need for environmental accounting.

16. Chatterjee and Finger 1993:3. See also Sachs 1993; Hawkins 1993.

17. For an argument that a more radical reformulation of the international relations literature is required see Saurin 1996.

18. For a succinct summary of the rulings, see Nordström and Vaughn, WTO Secretariat (1999), Annex 1, para. 140–147.

19. See Bernstein and Busumtwi-Sam (1998) for a further discussion of defining and understanding change at various levels of social structure.

20. There remains much debate on how radical a departure from the "Washington Consensus" the "Post-Washington Consensus" really entails. For example, Broad and Cavanagh argue that while financial elites within key governments and the international financial institutions seem willing to modify the goal of free capital flows, the consensus "largely holds with respect to trade policy" (1999:84). Similarly, Stiglitz's emphasis (1998, 1999) remains primarily on institutional support for markets to function smoothly and less on trade-offs. See also Naím 2000.

BIBLIOGRAPHY

Adams, W.M. 1990. *Green Development: Environment and Sustainability in the Third World*. London: Routledge.

Adler, Emanuel. 1990. "Cognitive Evolution: A Dynamic Approach for the Study of International Relations and Their Progress." In Emanuel Adler and Beverly Crawford, ed., *Progress in Postwar International Relations*, pp. 43–88. New York: Columbia University Press.

—— 1997. "Seizing the Middle Ground: Constructivism in World Politics." *European Journal of International Relations* 3:319–363.

—— and Peter M. Haas. 1992. "Conclusion: Epistemic Communities, World Order, and the Creation of a Reflective Research Program." *International Organization* 46:367–390.

The Agesta Group AB Sweden. 1982. *Twenty Years After Stockholm 1972–1992: A Report on the Implementation of the Stockholm Action Plan and on Priorities and Institutional Arrangements for the 1980s*. Berlin: Erich Schmidt Verlag.

Agrawala, Shardul. 1999. "Early Science-Policy Interactions in Climate Change: Lessons from the Advisory Group on Greenhouse Gases." *Global Environmental Change* 9:157–169.

AJIL. 1994. "Law of the Sea Forum: The 1994 Agreement on Implementation of the Seabed Provisions of the Convention on the Law of the Sea." *American Journal of International Law* 88:687–714.

Ambio. 1994. 23 1.

Andresen, Steinar and Willy Ostreng, eds. 1989. *International Resource Management: The Role of Science and Politics*. London: Belhaven Press.

Andresen, Steinar et al., eds. 2000. *Science and Politics in International Environmental Regimes: Between Integrity and Involvement.* Manchester: Manchester University Press.

Anderson, Terry and Donald R. Leal. 1991. *Free Market Environmentalism.* San Francisco: Pacific Research Institute for Public Policy, Westview Press.

Arda, Mehmet. 1996. "The United Nations Conference on Trade and Development." In Jacob Werksman, ed., *Greening International Institutions*, pp. 71–93. London: Earthscan.

Arizpe, L., R. Costanza, and W. Lutz. 1992. "Population and Natural Resource Use." In J.C.I. Dooge et al., eds., *An Agenda of Science for Environment and Development Into the 21st Century*, pp. 61–78. Cambridge: Cambridge University Press.

Ashley, Richard K. 1984. "The Poverty of Neorealism." *International Organization* 38:225–286.

Benedick, Richard Elliot. 1991. *Ozone Diplomacy: New Directions in Safeguarding the Planet.* Cambridge: Cambridge University Press.

Bennet, Colin. 1991. "What is Policy Convergence and What Causes It." *British Journal of Political Science* 21:215–233.

Bernstein, Steven and James Busumtwi-Sam. 1998. "The Social Structure of the Westphalian State System: Mapping Continuity and Change." Paper presented to the 39th Annual Convention of the International Studies Association, Minneapolis, 17–21 March.

Bernstein, Steven and Benjamin Cashore. 2000a. "Globalization, Four Paths of Internationalization and Domestic Policy Change: The Case of Ecoforestry in British Columbia, Canada." *Canadian Journal of Political Science* 33:67–99.

——— 2000b (forthcoming). "The International-Domestic Nexus: The Effects of International Trade and Environmental Politics on the Canadian Forest Sector." In Michael Howlett, ed., *Canadian Forest Policy: Regimes, Policy Dynamics and Institutional Adaptations.* Toronto: University of Toronto Press.

Bernstein, Steven, Richard Ned Lebow, Janice Gross Stein and Steve Weber. 2000. "God Gave Physics the Easy Problems: Adapting Social Science to an Unpredictable World." *European Journal of International Relations* 6:43–76.

Bhaskar, Roy. 1979. *The Possibility of Naturalism.* Brighten, U.K.: Harvester Press.

Biersteker, Thomas. 1992. "The 'Triumph' of Neoclassical Economics in the Developing World: Policy Convergence and Bases of Governance in the International Economic Order." In James Rosenau and Ernst-Otto Czempiel, eds., *Governance Without Government: Order and Change in World Politics*, pp. 102–131. Cambridge: Cambridge University Press.

——— and Cynthia Weber. 1996. *State Sovereignty as Social Construct.* Cambridge: Cambridge University Press.

Birnie, Patricia. 1992. "International Environmental Law." In Andrew Hurrell and Benedict Kingsbury, eds., *The International Politics of the Environment*, pp. 51–84. Oxford: Oxford University Press.

Blyth, Mark M. 1997. "Any More Bright Ideas? The Ideational Turn of Comparative Political Economy." *Comparative Politics* 29:229–250.

Bodansky, Daniel. 1991. "Scientific Uncertainty and the Precautionary Principle." *Environment* 33:4–5, 43–44.

———— 1994. "Prologue to the Climate Change Convention." In Irving M. Mintzer and J.A. Leonard, eds., *Negotiating Climate Change: The Inside Story of the Rio Convention*, pp. 45–74. Cambridge: Cambridge University Press.

Boehmer-Christiansen, Sonja. 1994a. "Global Climate Protection Policy: The Limits of Scientific Advice, Part 1." *Global Environmental Change* 4:140–159.

———— 1994b. "Global Climate Protection Policy: The Limits of Scientific Advice, Part 2." *Global Environmental Change* 4:185–200.

———— 1994c. "Scientific Uncertainty and Power Politics: The Framework Convention on Climate Change and the Role of Scientific Advice." In Bertram I. Spector et al., eds., *Negotiating International Regimes: Lessons Learned from the United Nations Conference on Environment and Development (UNCED)*, pp. 181–198. London: Graham and Trotman/Martinus Nijhoff.

Bolin, Bert. February 1994. "Politicians Deal With Politics, Scientists Better Not Get Mixed Up With That." Interview in *Change* 18:4–7.

————, B. Döös, J. Jäger and R. Warrick eds. 1986. SCOPE 29: *The Greenhouse Effect: Climate Change and Ecosystems*. Chichester: Wiley and Sons.

Bowler, Peter J. 1992. *The Norton History of the Environmental Sciences*. New York: Norton.

Boyle, Alan E. 1996. "The Rio Convention on Biological Diversity." In Michael Bowman and Catherine Redgwell, eds., *International Law and the Conservation of Biological Diversity*, pp. 33–49. London: Kluwer Law International.

Boyle, Robert H. 1999. "You're Getting Warmer . . . " *Audobon* (Jan/Feb). Available at http://www.magazine.audubon.org/global.html.

Braadbaart, Okke. 1998. "American Bias in Environmental Economics: Industrial Pollution Abatement and 'Incentives versus Regulations.' " *Environmental Politics* 7(2):134–152.

The Brandt Commission. 1980. *North-South: A Programme for Survival*. London: Pan Books.

———— 1983. *Common Crisis: North-South: Co-operation for World Recovery*. London: Pan Books.

Brenton, Tony. 1994. *The Greening of Machiavelli: The Evolution of International Environmental Politics*. London: Earthscan.

Broad, Robin and John Cavanagh. 1999. "The Death of the Washington Consensus?" *World Policy Journal* (Fall):79–88.

Brodeur, Paul. 1986. "Annals of Chemistry: In the Face of Doubt." *The New Yorker* (June 9):71–87.

Brooks, David B. 1992. "The Challenge of Sustainability: Is Integrating Environment and Economy Enough." *Policy Sciences* 25:401–408.

Brown, Donald A. 1998. "Making CSD Work." *Linkages* 3 2 (Available at www.iisd.ca/linkages/journal).

Bunge, Mario. 1991. "A Critical Examination of the New Sociology of Science." *Philosophy of the Social Sciences* 21:524–560.

Busumtwi-Sam, James. 1995. "The Role of the IMF and the World Bank in International Development." In Eric Fawcett and H. Newcombe, eds., *United Nations Reform*, pp. 248–266. Toronto: Dundurn Press.

———— and Steven Bernstein. 1997. "Institutionalization in World Politics." Unpublished paper. Simon Fraser University and University of Toronto.

Caldwell, Lynton Keith. 1990. *International Environmental Policy: Emergence and Dimensions*, 2nd ed. Durham and London: Duke University Press.

Campiglio, Luigi, et al., eds. 1994. *The Environment After Rio: International Law and Economics*. London: Graham and Trotman/Martinus Nijhoff.

Canadian Global Change Program. 1996. *Canadian Involvement in International Global Change Activities: A Compendium*. Ottawa: Canadian Global Change Program.

Cancian, Francesca. 1975. *What are Norms?* Cambridge: Cambridge University Press.

Caponera, Dante A. 1972. "Towards a New Methodological Approach in Environmental Law." *Natural Resources Journal* 12:133–152.

Caporaso, James. A. 1993. "International Relations Theory and Multilateralism: The Search for Foundations." In John G. Ruggie, ed. *Multilateralism Matters*, pp. 51–90. New York: Columbia University Press.

Carson, Rachel. 1962. *Silent Spring*. Greenwich, Conn.: Fawcett Books.

Chasek, Pamela. 1994a. "The Negotiating System of Environment and Development." In Bertram I. Spector et al., eds., *Negotiating International Regimes: Lessons Learned from the United Nations Conference on Environment and Development (UNCED)*, pp. 21–41. London: Graham and Trotman/Martinus Nijhoff.

———— 1994b. "The Story of the UNCED Process." In Bertram I. Spector et al., eds., *Negotiating International Regimes: Lessons Learned from the United Nations Conference on Environment and Development (UNCED)*, pp. 45–61. London: Graham and Trotman/Martinus Nijhoff.

Chatterjee, Pratap and Matthias Finger. 1994. *The Earth Brokers*. New York: Routledge.

Checkel, Jeff. 1993. "Ideas, Institutions, and the Gorbachev Foreign Policy Revolution." *World Politics* 45:271–300.

Chinkin, C.M. 1989. "The Challenge of Soft Law: Development and Change in International Law." *International and Comparative Law Quarterly* 38:850–866.

Chopra, Jarat and Thomas G. Weiss. 1995. "Sovereignty Under Siege." In Gene M. Lyons and M. Mastanduno, eds., *Beyond Westphalia? State Sovereignty and International Intervention*, pp. 91–114. Baltimore: Johns Hopkins University Press.

Clapp, Jennifer. 1994. "The Toxic Waste Trade With Less-Industrialized Countries: Economic Linkages and Political Alliances." *Third World Quarterly* 15:505–518.

———— 1998. "The Privatization of Global environmental governance: ISO 14000 and the Developing World." *Global Governance* 4:295–316.

Clark, William C. and R.E. Munn, eds. 1986. *Sustainable Development of the Biosphere*. Cambridge: Cambridge University Press.

Cohen, Steward et al. 1998. "Climate Change and Sustainable Development: Towards Dialogue." *Global Environmental Change* 8:341–371.

Colby, Michael E. 1990. *Environmental Management in Development: The Evolution of Paradigms* (World Bank Discussion Papers, 80). Washington, D.C.: The World Bank.

Colorado Journal of International Law and Policy. 1993. Special Issue, "United Nations Conference on Environment and Development." *Colorado Journal of International Law and Policy* 4 1 (Winter).

Commission of the European Communities. 1993. *Towards Sustainability: A European Community Programme of Policy and Action in Relation to the Environment and Sustainable Development*. Luxembourg: Office for Official Publications of the European Communities.

Conca, Ken. 1993. "Environmental Change and the Deep Structure of World Politics." In Ken Conca and Ronnie D. Lipschutz, eds., *The State and Social Power in Global Environmental Politics*, pp. 306–326. New York: Columbia University Press.

Costanza, Robert and Laura Cornwell. 1992. "The 4P Approach to Dealing with Scientific Uncertainty." *Environment* 34(9):8–20, 42.

Cox, Robert W. 1983. "Gramsci, Hegemony and International Relations: An Essay in Method." *Millennium: Journal of International Studies* 12:60–175.

—— 1986. "Social Forces, States and World Orders: Beyond International Relations Theory." In Robert O. Keohane, ed., *Neorealism and Its Critics*, pp. 204–254. New York: Columbia University Press.

—— 1987. *Production, Power, and World Order: Social Forces in the Making of History*. New York: Columbia University Press.

—— 1992. *Globalization, Multilateralism and Democracy* (the John W. Holmes Memorial Lecture). Providence: ACUNS Reports and Papers 1992–2.

Dabholkar, Uttam. 1989. "Environmental Perspective to the Year 2000 and Beyond: A Framework for Development." *Environmental Conservation* 16:49–53.

Dales, John Harkness. 1968. *Pollution, Property and Prices*. Toronto: University of Toronto Press.

de Almeida, Miguel Ozorio. 1972. "The Confrontation Between Problems of Development and Environment." In *Founex Report*, pp. 37–56. New York: Carnegie Endowment for International Peace.

Dessler, David. 1989. "What's At Stake in the Agent-Structure Debate?" *International Organization* 43:441–474.

Deudney, Daniel. 1993. "Global Environmental Rescue and the Emergence of World Domestic Politics." In Ken Conca and Ronnie D. Lipschutz, eds., *The State and Social Power in Global Environmental Politics*, pp. 280–305. New York: Columbia University Press.

Deutz, Andrew. 1997. "Institutionalized Review Processes of the Adequacy of Commitments in International Environmental Agreements." Ph.D. Dissertation, The Fletcher School of Law and Diplomacy.

Doherty, Ann. 1994. "The Role of Nongovernmental Organizations in UNCED." In Bertram I. Spector et al. eds., *Negotiating International Regimes: Lessons Learned from the United Nations Conference on Environment and Development (UNCED)*, pp. 199–218. London: Graham and Trotman/Martinus Nijhoff.

Dooge, J.C.I. et al., eds. 1992. *An Agenda of Science for Environment and Development into the 21st Century*. Cambridge: Cambridge University Press.

Döös, Bo R. 1991. *Environmental Issues Requiring International Action*. Laxenburg, Austria: International Institute for Applied Systems Analysis, RR-91–16.

Doty, Roxanne. 1997. "Aporia: A Critical Exploration of the Agent-Structure Problematique in International Relations Theory." *European Journal of International Relations* 3:365–392.

Dunlap, Riley E., George H. Gallup, Jr., and Alec M. Gallup. 1993. "Of Global Concern: Results of the Health of the Planet Survey." *Environment* 35(9):7–15, 33–39.

Dupuy, Pierre-Marie. 1991. "Soft Law and the International Law of the Environment." *Michigan Journal of International Law* 12:420–435.

Eckersley, Robyn. 1993. "Free Market Environmentalism: Friend of Foe?" *Environmental Politics* 2:1–19.

The Economist. 1987. "Making Sense of Brundtland." *The Economist* (May 2):15–16.

——— 1988. "The Greening of the Invisible Hand." *The Economist* (December 24):107–108.

——— 1991. "More Bright Ideas." *The Economist* (May 11):24–25.

——— 1992. "The Question Rio Forgets." *The Economist* (May 30):11–12.

El-Ashry, Mohamed T. 1993. "Development Assistance Institutions and Sustainable Development." *The Washington Quarterly* 16:83–96.

Elster, Jon. 1983. *Explaining Technical Change: A Case Study in the Philosophy of Science*. Cambridge: Cambridge University Press.

Engfeldt, Lars-Goran. 1973. "The United Nations and the Human Environment — Some Experiences." *International Organization* 27:393–412.

Environmental News Service (ENS). 1997. "The UN Gets a Wake-up Call." (June 29), made available through the EnviroLink News Service http://www.newsdesk@envirolink.org

——— 1998. "Shell Adopts 10% Emissions Cut." (October 17).

Fairman, David. 1994. "Report of the Independent Evaluation of the Global Environment Facility Pilot Phase" and "Commentary: The New GEF." *Environment* 36(6):25–30, 37–39.

Falk, Richard. 1995. "Liberalism at the Global Level: The Last of the Independent Commissions?" *Millennium* 24:563–576.

FAO. 1983. "International Undertaking on Plant Genetic Resources of the FAO." Reproduced in *Basic Documents in International Environmental Law* vol. 1, pp. 113–18. London: Graham and Trotman, 1992.

Feketekuty, Geza. 1993. "The Link Between Trade and Environmental Policy." *Minnesota Journal of Global Trade* 2:171–205.

Financial Times. 2000. "Shell: Oil Group Launches International Carbon Emission Market" January 27, 2000, from climate-L newslist (February 6).

Finger, Matthias. 1993. "Politics of the UNCED Process." In Wolfgang Sachs, ed., *Global Ecology: A New Arena of Political Conflict*, pp. 36–48. London: Zed Books.

Finklestein, Lawrence S. 1995. "What is Global Governance." *Global Governance* 1:367–371.

Finnemore, Martha. 1993. "International Organizations as Teachers of Norms: The United Nations Educational, Scientific, and Cultural Organization and Science Policy." *International Organization* 47:565–598.

——— 1996a. *National Interests in International Society*. Ithaca: Cornell University Press.

——— 1996b. "Constructing Norms of Humanitarian Intervention." In Peter J. Katzenstein, ed., *The Culture of National Security*, pp. 153–85. New York: Columbia University Press.

——— and Kathryn Sikkink. 1998. "International Norm Dynamics and Political Change." *International Organization* 52:887–917.

Florini, Ann. 1996. "The Evolution of International Norms." *International Studies Quarterly* 40:363–389.

Founex Report on Development and Environment. 1972. Reprinted in *Environment and Development: The Founex Report on Development and Environment, with commentaries by Miguel Ozorio de Almeida, Wilfred Beckerman, Ignacy Sachs, and Gamani Corea.* New York: Carnegie Endowment for International Peace.

Franck, Thomas. 1990. *The Power of Legitimacy Among Nations.* New York: Oxford University Press.

French, Hilary F. 1993. "The GATT: Menace or Ally." *Worldwatch* (Sept./Oct.): 12–19.

Frost, Mervyn. 1996. *International Relations: A Constitutive Theory.* Cambridge: Cambridge University Press.

Gale, Fred and Cheri Burda. 1997. "The Pitfalls and Potential of Eco-Certification as a Market Incentive for Sustainable Forest Management." In Chris Tollefson, ed., *The Wealth of Forests: Markets, Regulation and Sustainable Forestry*, pp. 414–441. Vancouver: University of British Columbia Press.

Garrett, Geoffrey and Barry R. Weingast. 1993. "Ideas, Interests, and Institutions: Constructing the European Community's Internal Market." In Judith Goldstein and Robert O. Keohane, eds., *Ideas and Foreign Policy: Beliefs, Institutions, and Political Change*, pp. 173–206. Ithaca: Cornell University Press.

Germain, Randall D. and Michael Kenny. 1998. "Engaging Gramsci: International Relations Theory and the New Gramscians." *Review of International Studies* 24:3–21.

Getches, David H. 1993. "Foreword: The Challenge of Rio." *Colorado Journal of International Environmental Law and Policy* 4:1–19.

Giddens, Anthony. 1979. *Central Problems in Social Theory.* London: Macmillan.

Gill, Stephen. 1989. *American Hegemony and the Trilateral Commission.* Cambridge: Cambridge University Press.

—— ed. 1993. *Gramsci, Historical Materialism and International Relations.* Cambridge: Cambridge University Press.

Gilpin, Robert. 1987. *The Political Economy of International Relations.* Princeton, NJ: Princeton University Press.

Goldstein, Judith and Robert O. Keohane, eds. 1993a. *Ideas and Foreign Policy: Beliefs, Institutions, and Political Change.* Ithaca: Cornell University Press.

—— 1993b. "Ideas and Foreign Policy: An Analytical Framework." In Judith Goldstein and Robert O. Keohane, eds., *Ideas and Foreign Policy: Beliefs, Institutions, and Political Change*, pp. 3–30. Ithaca: Cornell University Press.

Goodland, Robert J.A. 1992. "Environmental Priorities for Financing Institutions." *Environmental Conservation* 19:9–21.

——, Herman E. Daly, and Salah El Serafy. 1993. "The Urgent Need for Rapid Transition to Global Environmental Sustainability." *Environmental Conservation* 20:297–309.

Gorsuch, Anne. 1982. "Eight Principles for Sound Environmental Development." Speech to the Session of a Special Character of UNEP's Governing Council, reproduced in *Uniterra* 2:24–25.

Group of 77. 1981. *Collected Documents of the Group of 77.* New York. Oceana Publications.

Grove, Richard. 1991. "The First Environmentalists." *Wilson Quarterly* (Spring): 66–67.

Grubb, Michael. 1998. "Corrupting the Climate: Economic Theory and the Politics of International Environmental Agreements." Valedictory lecture given at Chatham House, Royal Institute of International Affairs on 20th October.

———— et al. 1993. *The Earth Summit Agreements*. London: Earthscan.

———— with Christiaan Vrolijk and Duncan Brack. 1999. *The Kyoto Protocol: A Guide and Assessment*. London: Royal Institute of International Affairs and Earthscan.

Haas, Peter M. 1989. "Do Regimes Matter? Epistemic Communities and Mediterranean Pollution Control." *International Organization* 43:377–403.

———— 1990. *Saving the Mediterranean: The Politics of International Environmental Protection*. New York: Columbia University Press.

———— 1992a. "Obtaining International Environmental Protection Through Epistemic Consensus." In Ian H. Rowlands and Malory Greene, eds., *Global Environmental Change and International Relations*, pp. 38–59. London: Macmillan.

———— ed. 1992b. "Power, Knowledge, and International Policy Coordination." Special Edition of *International Organization* 46(1).

———— 1992c. "Introduction: Epistemic Communities and International Policy Coordination." *International Organization* 46:1–35.

———— 1992d. "Banning Chlorofluorocarbons: Epistemic Community Efforts to Protect Stratospheric Ozone." *International Organization* 46:197–224.

———— 1996. "The Future of International Environmental Governance." Paper presented at the 37th Annual Convention of the International Studies Association, April 16–20, San Diego, California.

———— and Ernst B. Haas. 1995 "Learning to Learn: Improving International Governance." *Global Governance* 1:255–285.

————, Marc A. Levy, and Edward A Parson. 1992. "Appraising the Earth Summit: How Should We Judge UNCED's Success." *Environment* 34(8):6–11, 26–33.

————, Robert O. Keohane, and Marc A. Levy, eds. 1993. *Institutions for the Earth: Sources of Effective International Environmental Protection*. Cambridge: MIT Press.

Hagen, Joel B. 1992. *An Entangled Bank: The Origins of Ecosystem Ecology*. New Brunswick, NJ: Rutgers University Press.

Hahn, Robert and Robert Stavins. 1991. "Incentive-Based Environmental Regulation: A New Era from an Old Idea?" *Ecology Law Quarterly* 18:1–42.

———— 1992. "Economic Incentives for Environmental Protection: Integrating Theory and Practice." *The American Economics Review* 82:464–468.

Hajer, Maarten A. 1995. *The Politics of Environmental Discourse: Ecological Modernization and the Policy Process*. Oxford: Clarendon Press, 1995.

Hajnal, Peter I., ed. 1989. *The Seven Power Summit: Documents from the Summits of Industrialized Countries 1975–1989*. New York: Kraus International. Supplement, 1991.

Hajost, Scott A. 1994. "The Role of the United States." In Luigi Campiglio et al., eds., *The Environment After Rio: International Law and Economics*, pp. 15–21. London: Graham and Trotman/Martinus Nijhoff.

Hall, Peter A., ed. 1989a. *The Political Power of Economic Ideas*. Princeton, NJ: Princeton University Press.

———— 1989b. "Conclusion: The Politics of Keynesian Ideas." In *The Political Power of Economic Ideas*, 361–391. Princeton, NJ: Princeton University Press.

Hansen, Eric. 1998. "Certified Forest Products Market Place." In United Nations Timber Committee, ed., *Forest Products Annual Market Review*, pp. 17–28. Geneva, Switzerland: United Nations Timber Committee.

Hasenclever, Andreas, Peter Mayer, and Volker Rittberger. 1996. "Interests, Power, Knowledge: The Study of International Regimes." *Mershon International Studies Review* 40:177–228.

——— 1997. *Theories of International Regimes.* Cambridge: Cambridge University Press.

Hawkins, Ann. 1993. "Contested Ground: International Environmentalism and Global Climate Change." In Ken Conca and Ronnie D. Lipschutz, eds., *The State and Social Power in Global Environmental Politics*, pp. 221–245. New York: Columbia University Press.

Hayward, Tim. 1994. *Ecological Thought: An Introduction.* Cambridge: Polity Press.

"Heating up the Climate Change Debate." 1996. *Environment* 38:22–23.

Hecht, A. and D. Tirpak. 1995. "Framework Agreement on Climate Change: A Scientific and Policy History." *Climatic Change* 29:371–402.

Helleiner, Eric. 1996. "International Political Economy and the Greens." *New Political Economy* 1:59–77.

Herber, Bernard P. 1991. "The Common Heritage Principle: Antarctica and the Developing Nations." *American Journal of Economics and Sociology* 50:391–406.

Herter, Christian A. Jr. and Jill E. Binder. 1993. *The Role of the Secretariat in Multilateral Negotiation: The Case of Maurice Strong and the 1972 UN Conference on the Human Environment.* Washington, D.C.: The Johns Hopkins Foreign Policy Institute.

Hirschman, Albert O. 1989. "How the Keynesian Revolution Was Exported from the United States, and Other Comments." In Peter A. Hall, ed., *The Political Power of Economic Ideas*, pp. 347–359. Princeton, NJ: Princeton University Press.

Hollis, Michael and Steve Smith. 1990. *Explaining and Understanding International Relations.* Oxford: Clarendon Press.

Howarth, Richard B. and Patricia A. Monahan. 1996. "Economics, Ethics, and the Climate Policy: Framing the Debate." *Global and Planetary Change* 11:187–199.

Hull, Carrie. 1999. "When Something is to be Done: Proof of Environmental Harm and the Philosophical Tradition." *Environmental Values* 8:3–25.

Humphreys, David. 1996. "The Global Politics of Forest Conservation Since the UNCED." *Environmental Politics* 5:231–256.

Hunter, David B. and Lori Udall. 1994. "The World Bank's New Inspection Panel." *Environment* 36(9):2–3, 44–45.

Hurrell, Andrew and Benedict Kingsbury, eds. 1992. *The International Politics of the Environment.* Oxford: Oxford University Press.

ICSU. 1973. *ICSU Bulletin* 29 (December).

——— 1971. *ICSU Bulletin* 25 (December).

——— 1969. *ICSU Bulletin* 19 (November).

IDRC. 1993. *The Earth Summit CD-ROM.* Ottawa: International Development Research Centre on behalf of the United Nations.

Ikenberry, G. John and Charles A. Kupchan. 1990. "Socialization and Hegemonic Power." *International Organization* 17:283–315.

Imber, Mark F. 1994. *Environment, Security and UN Reform.* London: Macmillan.

Ingham, Alan. 1994. "The Market for Sulphur Dioxide Permits in the USA and UK." In Caroline Thomas, ed., *Rio: Unravelling the Consequences*, pp. 98–122. London: Frank Cass.

International Institute for Sustainable Development (IISD). 1997a. "Summary of the Nineteenth United Nations General Assembly Special Session to Review Implementation of Agenda 21." *Earth Negotiations Bulletin* 5(88). Available from the International Institute for Sustainable Development's website at www.iisd.ca.

———— 1997b. "Report of the Third Conference of the Parties to the Framework Convention on Climate Change." *Earth Negotiations Bulletin* 12(76).

———— 1998. "Report of the Fourth Conference of the Parties to the UN Framework Convention on Climate Change." *Earth Negotiations Bulletin* 12(97).

———— 1999. "Summary of the Fifth Conference of the Parties to the Basel Convention on the Control of Transboundary Movements of Hazardous Wastes and Their Disposal: 6–10 December." *Earth Negotiations Bulletin* 20(6).

———— 2000a. "Report of the Resumed Session of the Extraordinary Meeting of the Conference of the Parties for the Adoption of the Protocol on Biosafety to the Convention on Biological Diversity: 24–28 January 2000." *Earth Negotiations Bulletin* 9(137).

———— 2000b. "Summary of the Fourth Session of the Intergovernmental Forum on Forests: January 31—February 11 2000." *Earth Negotiations Bulletin* 13(66).

———— 2000c. "Summary of the Eighth Session of the UN Commission on Sustainable Development: April 24–May 5, 2000." *Earth Negotiations Bulletin* 5(157).

———— 2000d. "Second World Bank European Sustainable Development Forum 5–6 June 2000." *Sustainable Developments* 37(1).

———— 2000e. Summary of the Sixth Conference of the Parties to the Framework Convention on Climate Change: 13–25 November 2000." *Earth Negotiations Bulletin* 12(163).

International Legal Materials. Various years.

International Studies Quarterly. 1996. 40(3).

International Union for Conservation of Nature and Natural Resources (IUCN). 1980. *World Conservation Strategy: Living Resource Conservation for Sustainable Development.* Gland, Switzerland: IUCN.

IPCC. 1995. *Second Assessment Report* (Synthesis of Scientific-Technical Information Relevant to Interpreting Article 2 of the UN Framework Convention on Climate Change 1995, and executive summaries of Working Groups I, II, and III).

IPCC. 2001. "Summary for Policymakers." *Working Group 1 Third Assessment Report.* Shanghai Draft, January 21.

Jackson, Robert H. 1990. *Quasi-states: Sovereignty, International Relations and the Third World.* Cambridge: Cambridge University Press.

Jepperson, Ronald L., Alexander Wendt and Peter J. Katzenstein. 1996. "Norms, Identity, and Culture in National Security." In Peter J. Katzenstein, ed., *The Culture of National Security,* pp. 33–75. New York: Columbia University Press.

Jordan, Andrew. 1994a. "Financing the UNCED Agenda: The Controversy Over Additionality." *Environment* 36(3):16–20, 26–34.

———— 1994b. "Paying the Incremental Costs of Global Environmental Protection: The Evolving Role of GEF." *Environment* 36(6):12–20, 31–36.

Katzenstein, Peter J, ed. 1996. *The Culture of National Security: Norms and Identity in World Politics.* New York: Columbia University Press.

Keck, Margaret and Kathryn Sikkink. 1998. *Activists Beyond Borders: Transnational Issue Networks in International Politics.* New York: Cornell University Press.

Kell, George and John G. Ruggie. 1999. "Global Markets and Social Legitimacy: The Case of the 'Global Compact.'" Paper presented at an international conference: Governing the Public Domain Beyond the Era of the Washington Consensus? Redrawing the Line Between the State and the Market, York University, Toronto, Canada, November 4–6.

Kellermann, Henry J. 1973. "Stockholm and the Role of Science." *Bioscience* 23:485–487.

Kennedy, Kevin C. 1994. "Reforming U.S. Trade Policy to Protect the Global Environment: A Multilateral Approach." *Harvard Environmental Law Review* 18:185–234.

Keohane, Robert O. 1978 "Economics, Inflation, and the Role of the State: Political Implications of the McCracken Report." *World Politics* 31:108–128.

—— 1984. *After Hegemony*. Princeton: Princeton University Press.

—— 1988. "International Institutions: Two Approaches." *International Studies Quarterly* 32:379–396.

—— 1989. *International Institutions and State Power*. Boulder: Westview Press.

—— 1991. "The Theory of Hegemonic Stability and Changes in International Economic Regimes, 1967–1977." Reprinted in George T. Crane and Abla Amawi, eds., *The Theoretical Evolution of International Political Economy*, pp. 245–262. New York: Oxford University Press.

—— 1993. "Institutional Theory and the Realist Challenge After the Cold War." In David A. Baldwin, ed., *Neorealism and Neoliberalism*, pp. 269–300. New York: Columbia University Press.

Kiekens, Jean-Pierre. 1997. *Certification: International Trends and Forestry and Trade Implications*. Bruxelles: Environmental Strategies Europe.

King, Gary, Robert O. Keohane, and Sydney Verba. 1994. *Designing Social Inquiry: Scientific Inference in Qualitative Research*. Princeton, NJ: Princeton University Press.

Kiss, Alexandre. 1994. "The Rio Declaration on Environment and Development." In Luigi Campiglio et al., ed., *The Environment After Rio: International Law and Economics*, pp. 55–64. London: Graham and Trotman/Martinus Nijhoff.

Klotz, Audie. 1995. *Protesting Prejudice: Apartheid and the Politics of Norms in International Relations*. Ithaca: Cornell University Press.

Kocs, Stephen A. 1994. "Explaining the Strategic Behavior of States: International Law as System Structure." *International Studies Quarterly* 38:535–56.

Kovar, Jeffrey D. 1993 "A Short Guide to the Rio Declaration." *Colorado Journal of International Environmental Law and Policy* 4:119–140.

Kowalok, Michael E. 1993. "Research Lessons from Acid Rain, Ozone Depletion, and Global Warming." *Environment* 35(6):13–20, 35–38.

Kowert, Paul and Jeffrey Legro. 1996. "Norms, Identity, and Their Limits: A Theoretical Reprise." In Peter J. Katzenstein, ed., *The Culture of National Security*, pp. 451–497. New York: Columbia University Press.

Kraft, Michael E. and Norman J. Vig. 1984. "Environmental Policy in the Reagan Presidency." *Political Science Quarterly* 99:415–439.

—— 1994. *Environmental Policy in the 1990s*. 2nd ed. Washington, D.C.: Congressional Quarterly Press.

Krasner, Stephen D., ed. 1983. *International Regimes*. Ithaca: Cornell University Press.

———— 1985. *Structural Conflict: The Third World Against Global Liberalism*. Berkeley: University of California Press.

———— 1988. "Sovereignty: An Institutional Perspective." *Comparative Political Studies* 21:66–94.

Kratochwil, Friedrich. 1989. *Rules, Norms, Decisions: On the Conditions of Practical and Legal Reasoning in International Relations and Domestic Affairs*. Cambridge: Cambridge University Press.

———— 1995. "Sovereignty as Dominium: Is there a Right of Humanitarian Intervention?" In Gene M. Lyons and Michael Mastanduno, eds., *Beyond Westphalia? State Sovereignty and International Intervention*, pp. 21–42. Baltimore and London: Johns Hopkins University Press.

———— and John Gerard Ruggie. 1986. "International Organization: A State of the Art on an Art of the State." *International Organization* 40:753–775.

Krueger, Jonathon. 1999. *International Trade and the Basel Convention*. London: Earthscan.

Krugman, Paul. 1995. "Dutch Tulips and Emerging Markets." *Foreign Affairs* (July/August):28–44.

Lemons, John and Donald A. Brown. 1995. "The Role of Science in Sustainable Development and Environmental Protection Decisionmaking." In John Lemons and Donald A. Brown, *Sustainable Development: Science, Ethics, and Public Policy*, pp. 11–38. Dordrecht: Kluwer Academic Publishers.

Levy, Mark A, Oran R. Young, and Michael Zürn. 1995. "The Study of International Regimes." *European Journal of International Relations* 1:267–330.

Lewis, Martin W. 1992. *Green Delusions: An Environmentalist Critique of Radical Environmentalism*. Durham, N.C.: Duke University Press.

Lipschutz, Ronnie D. and Ken Conca, eds. 1993. *The State and Social Power in Global Environmental Politics*. New York: Columbia University Press.

Litfin, Karen. 1993. "Eco-regimes: Playing Tug of War with the Nation-State." In Ken Conca and Ronnie D. Lipschutz, eds., *The State and Social Power in Global Environmental Politics*, pp. 94–117. New York: Columbia University Press.

———— 1994. *Ozone Discourses*. New York: Columbia University Press.

MacNeill, Jim. 1971. *Environmental Management*. Ottawa. Information Canada.

———— 1984. *Mandate: Key Issues, Strategy and Workplan*. Mimeo from author.

———— Pieter Winsemius, and Taizo Yakushiji. 1991. *Beyond Interdependence: The Meshing of the World's Economy and the Earth's Ecology*. New York: Oxford University Press, Trilateral Commission.

Majone, Giandomenico. 1989. *Evidence, Argument and Persuasion in the Policy Process*. New Haven, London: Yale University Press.

March, James G. and Johan P. Olsen. 1984. "The New Institutionalism: Organizational Factors in Political Life." *American Political Science Review* 78:734–749.

———— 1989. *Rediscovering Institutions: The Organizational Basis of Politics*. New York: Free Press.

———— 1998. "The Institutional Dynamics of International Order." *International Organization* 52:943–970.

Marshall, Will and Martin Schram, eds. 1993. *Mandate for Change*. New York: Berkley Books for the Progressive Policy Institute and the Democratic Leadership Council.

Marton-Lefèvre, Julia. 1994. "The Role of the Scientific Community in the Preparation of and Follow-up to UNCED." In Bertram I. Spector et al., eds., *Negotiating International Regimes: Lessons Learned from the United Nations Conference on Environment and Development (UNCED)*, pp. 171–180. London: Graham and Trotman/Martinus Nijhoff.

McConnell, Fiona. 1996. *The Biodiversity Convention: A Negotiating History.* London: Kluwer Law International.

McCormick, John. 1989. *Reclaiming Paradise: The Global Environmental Movement.* Bloomington: Indiana University Press.

McKeown, Timothy J. 1999. "Case Studies and the Statistical Worldview: Review of King, Keohane, and Verba's Designing Social Inquiry: Scientific Inference in Qualitative Research." *International Organization* 53:161-190.

McNamara, Kathleen R. 1998. *The Currency of Ideas: Monetary Politics in the European Union.* Ithaca, NY: Cornell University Press.

Meadows, Donella H., et al. 1972. *The Limits to Growth.* 2nd ed. New York: Signet.

Mensah, Chris K. 1994. "The Role of Developing Countries." In Luigi Campiglio et al., eds., *The Environment After Rio: International Law and Economics*, pp. 33–52. London: Graham and Trotman/Martinus Nijhoff.

Meyer, John W. et al. 1997. "The Structuring of a World Environmental Regime, 1870–1990." *International Organization* 51:623–51.

Miller, Richard W. 1987. *Fact and Method: Explanation, Confirmation, and Reality in the Natural and the Social Sciences.* Princeton, NJ: Princeton University Press.

Mintzer, Irving M. and J.A. Leonard, eds. 1994. *Negotiating Climate Change: The Inside Story of the Rio Convention.* Cambridge: Cambridge University Press.

Modelski, George and Kazimerz Poznanski. 1996. "Evolutionary Paradigms in the Social Sciences." *International Studies Quarterly* 40:315–319.

Moffatt, Ian. 1996. *Sustainable Development: Principles, Analysis and Policies.* New York: Parthenon.

Molina, M.J. and F.S. Rowland. 1974. "Stratospheric Sink for Chlorofluoromethanes: Chlorine Atom-Catalysed Destruction of Ozone." *Nature* 249:810–812.

Moravcsik, Andrew. 1997. "A Liberal Theory of International Politics." *International Organization* 51:513–553.

Munn, R.E. 1987. *Environmental Prospects for the Next Century: Implications for Long-Term Policy and Research Strategies.* RR-87–15. Laxenburg, Austria: IIASA.

——— 1992. "Towards Sustainable Development." *Atmospheric Environment* 26A:2725–2731.

Munton, Don. 1998. "Dispelling the Myths of the Acid Rain Story." *Environment* 40(6):4–7, 27–34.

Murphy, Alexander B. 1996. "The Sovereign State System as Political-Territorial Ideal: Historical and Contemporary Considerations." In Thomas J. Biersteker and Cynthia Weber, eds., *State Sovereignty as Social Construct*, pp.81–120. Cambridge: Cambridge University Press.

Murphy, Craig N. 1998. "Understanding IR: Understanding Gramsci." *Review of International Studies* 24:417–425.

Nadelmann, Ethan. 1990. "Global Prohibition Regimes: The Evolution of Norms in International Society." *International Organization* 44:479–526.

Naím, Moisés. 2000. "Washington Consensus or Washington Confusion?" *Foreign Policy* 118:86–103.

New York Times. 1997. "World Forest Talks End in Division on Whether to Curb Logging." *New York Times,* February 22, p. 7.

Nordström, Häkan and Scott Vaughan, WTO Secretariat. 1999. *Trade and Environment* (Special Studies #4). Geneva: World Trade Organization.

Odum, Eugene P. 1953, 1959, 1971. *Fundamentals of Ecology.* Philadelphia: Saunders. The 1953 edition is referenced in this work.

Odum, Howard. 1971. *Environment, Power, and Society.* New York: John Wiley.

OECD. 1971. Environment Committee. *Problems and Instruments Relating to the Allocation of Environmental Costs* (prepared for UNHCE). Paris: OECD, 25 Aug.

——— 1972. "Guiding Principles on the Environment." *International Legal Materials* 11:1172.

——— 1973a. *OECD: History. Aims. Structure* 2nd ed. Paris: OECD.

——— 1973b. *OECD at Work for Environment* 2nd ed. Paris: OECD.

——— 1974. "OECD Council Recommendation on the Implementation of the Polluter-Pays Principle." (November 21). Reproduced in Edith Brown Weiss, Daniel Barstow Magraw, and Szasz, Paul C. 1992. *International Environmental Law: Basic Instruments and References.* Dobbs Ferry, NY: Transnational Publishers.

——— 1975. *The Polluter Pays Principle.* Paris: OECD.

——— 1985. *Environment and Economics: Results of the International Conference on Environment and Economics, 18th-21st June 1984.* Paris: OECD.

——— 1994a. *Managing the Environment: The Role of Economic Instruments.* Paris: OECD.

——— 1994b. *Project and Policy Appraisal: Integrating Economics and the Environment.* Paris: OECD.

——— 1995. *Global Warming: Economic Dimensions and Policy Responses.* Paris: OECD.

——— 1998. *Lessons from Existing Trading Systems for International Greenhouse Gas Emission Trading.* Paris: OECD.

Onuf, Nicholas G. 1997. "How Things Get Normative." Revised version of a paper presented to conference on "International Norms." Hebrew University of Jerusalem, 26–May 27.

O'Riordan, Timothy. 1995a. "Frameworks for Choice: Core Beliefs and the Environment." *Environment* 37(8):4–9, 25–29.

——— ed. 1995b. *Environmental Science for Environmental Management.* Essex: Longman Scientific and Technical.

——— and James Cameron, eds. 1994. *Interpreting the Precautionary Principle.* London: Earthscan.

Oye, Kenneth, ed. 1986. *Cooperation Under Anarchy.* Princeton: Princeton University Press.

Pallemaerts, Marc. 1994. "International Environmental Law from Stockholm to Rio: Back to the Future?" In Philippe Sands, ed., *Greening International Law,* pp. 1–19. New York: The New Press.

——— 1996. "International Environmental Law in the Age of Sustainable Development: A Critical Assessment of the UNCED Process." *Journal of Law and Commerce* 15:623–676.

Parson, Edward A. 1993. "Protecting the Ozone Layer: The Evolution and Impact of International Institutions." In Peter M. Haas, Robert O. Keohane, and Marc A. Levy, eds., *Institutions for the Earth: Sources of Effective International Environmental Protection*, pp. 27–73. Cambridge: MIT Press.

———— and Owen Greene. 1995. "The Complex Chemistry of the International Ozone Agreements." *Environment* 37(2):16–20, 35–43.

———— Peter M. Haas, and Marc A. Levy. 1992. "A Summary of the Major Documents Signed at the Earth Summit and the Global Forum." *Environment* 34(8): 12–15, 34–36.

Passell, Peter. 1988. "Private Incentives as Pollution Curb." *New York Times* (October 19):D2.

Paterson, Matthew. 1996. *Global Warming and Global Politics*. London: Routledge.

Pauly, Louis W. 1997. *Who Elected the Bankers? Surveillance and Control in the World Economy*. Ithaca: Cornell University Press.

Payoyo, Peter Bautista. 1997. *Cries of the Sea: World Inequality, Sustainable Development and the Common Heritage of Humanity*. The Hague: Martinus Nijhoff.

Pearce, David, Anil Markandya and Edward B. Barbier. 1989. *Blueprint for a Green Economy*. London: Earthscan.

Pearce, David, ed. 1991. *Blueprint 2: Greening the World Economy*. London: Earthscan.

Pearce, David W. and Jeremy J. Warford. *World Without End: Economics, Environment, and Sustainable Development*. 1993. Oxford: Oxford University Press for the World Bank.

Pearson, Charles S. 1994. "Testing the System: GATT + PPP = ?" *Cornell International Law Journal* 27:353–575.

Piddington, Kenneth. 1992. "The Role of the World Bank." In Andrew Hurrell and Benedict Kingsbury, eds., *The International Politics of the Environment*, pp. 212–227. Oxford: Oxford University Press.

Pigou, Arthur Cecil. 1920. *The Economics of Welfare*. London: Macmillan.

Polkinghorne, Donald. 1988. *Narrative Knowing and the Human Sciences*. Albany: State University of New York Press.

Porras, Ileana. 1994. "The Rio Declaration: A New Basis for International Cooperation." In Philippe Sands, ed., *Greening International Law*, pp. 20–33. New York: The New Press.

Porter, Gareth and Janet Welsh Brown. 1991. *Global Environmental Politics*. Boulder, Colo.: Westview Press.

———— 1996. *Global Environmental Politics*. 2nd edition. Boulder, CO: Westview Press.

Princen, Thomas and Matthias Finger. 1994. *Environmental NGOs in World Politics: Linking the Local and the Global*. London: Routledge.

Project 88: Harnessing Market Forces to Protect Our Environment: Initiatives for the New President. 1988. A public policy study sponsored by Senators Timothy E. Wirth and John Heinz. Washington D.C.: Project 88.

Project 88 — Round II, Incentives for Action: Designing Market-Based Environmental Strategies. 1991. A public policy study sponsored by Senator Timothy E. Wirth and Senator John Heinz. Washington D.C.

Rana, Harminderpal Singh. 1994. "The 'Common Heritage of Mankind' and The Final Frontier." *Rutgers Law Journal* 26:225–250.

Raymond, Gregory A. 1997. "Problems and Prospects in the Study of International Norms." *Mershon International Studies Review* 41:205–245.

"Recommendations from Sigma Xi and ASCEND 21." 1992. *Environment* 34(8): 5, 40.

Reitan, Marit. 1998. "Ecological Modernisation and 'Realpolitik': Ideas, Interests and Institutions." *Environmental Politics* 7:1–26.

Reiterer, Michael. 1997. "Trade and Environment: Reflections on the Impact of the OECD Joint Session." *International Environmental Affairs* 9:69–81.

Reus-Smit, Christian. 1997. "The Constitutional Structure of International Society: The Nature of Fundamental Institutions." International Organization 51: 555–89.

Reuters. 1997. "Economists Say Climate Change Must Be Halted." February 14.

Rich, Bruce. 1990. "The Emperor's New Clothes: The World Bank and Environmental Reform." *World Policy Journal* 7:205–329.

—— 1994. *Mortgaging the Earth: The World Bank, Environmental Impoverishment and the Crisis of Development.* Boston: Beacon Press.

Ricupero, Rubens. 1993. "UNCED and Agenda 21: Chronicle of a Negotiation." *Colorado Journal of International Environemntal Law and Policy* 4:81–101.

Riddell, Robert. 1981. *Ecodevelopment.* Westmead, England: Gower.

Risse-Kappen, Thomas. 1994. "Ideas Do Not Float Freely: Transnational Coalitions, Domestic Structures, and the End of the Cold War." *International Organization* 48:185–214.

—— ed. 1995. *Bringing Transnational Relations Back In: Non-State Actors, Domestic Structures, and International Institutions.* Cambridge: Cambridge University Press.

Risse, Thomas, Stephen C. Ropp, and Kathryn Sikkink, eds. 1999. *The Power of Human Rights: International Norms and Domestic Change.* Cambridge: Cambridge University Press.

Rodrik, Dani. 1994. "The Rush to Free Trade in the Developing World: Why So Late? Why Now? Will It Last?" In Stephan Haggard and Steven B. Webb, eds., *Voting for Reform: Democracy, Political Liberalization, and Economic Adjustment,* pp. 61–88. Washington, DC: World Bank.

Rogers, Adam. 1993. *The Earth Summit: A Planetary Reckoning.* Los Angeles: Global View Press.

Rosenau, James and Ernst-Otto Czempiel eds. 1992. *Governance Without Government: Order and Change in World Politics.* Cambridge: Cambridge University Press.

Rothstein, Robert. 1979. *Global Bargaining: UNCTAD and the Quest for a New International Economic Order.* Princeton, NJ: Princeton University Press.

—— 1984. "Consensual Knowledge and International Collaboration: Some Lessons from the Commodity Negotiations." *International Organization* 38: 733–762.

Rowland, F. Sherwood. 1993. "President's Lecture: The Need for Scientific Communication with the Public." *Science* 260:1571–1576.

Rowland, Wade. 1973. *The Plot to Save The World.* Toronto: Clarke, Irwin and Company ltd.

Rowlands, Ian. 1995a. *The Politics of Global Atmospheric Change.* Manchester: Manchester University Press.

———— 1995b. "The Climate Change Negotiations: Berlin and Beyond." *Journal of Environment and Development* 4:146–163.

Ruggie, John Gerard. 1975. "International Responses to Technology: Concepts and Trends." *International Organization* 29:557–583.

———— 1983. "International Regimes, Transactions, and Change: Embedded Liberalism in the Postwar Economic Order." In Stephen D. Krasner, ed., *International Regimes*, pp. 195–232. Ithaca: Cornell University Press.

————1986. "Continuity and Transformation in the World Polity." In Robert O. Keohane, ed., *Neorealism and its Critics*, pp. 131–57. New York: Columbia University Press.

———— 1991. "Embedded Liberalism Revisited: Institutions and Progress in International Economic Relations." In Emanuel Adler and Beverly Crawford, eds., *Progress in Postwar International Relations*, pp. 201–234. New York: Columbia University Press.

———— 1992. "Multilateralism: The Anatomy of an Institution." *International Organization* 46:561–598.

———— 1993. "Territoriality and Beyond: Problematizing Modernity in International Relations." *International Organization* 47:139–174.

———— 1995. "Peace in Our Time? Causality, Social Facts and Narrative Knowing." *American Society of International Law. Proceedings of the 89th Annual Meeting: Structures of World Order* 82:93–100.

———— 1998. *Constructing the World Polity.* London: Routledge.

Rupert, Mark. 1998. "(Re-)Engaging Gramsci: A Response to Germain and Kenny." *Review of International Studies* 24:427–434.

Sabatier, Paul and Hank Jenkins-Smith. 1993. "The Advocacy Coalition Framework: Assessment, Revisions, and Implications for Scholars and Practitioners." In Paul Sabatier and Hank Jenkins-Smith, eds., *Policy Change and Learning: An Advocacy Coalition Approach*, pp. 211–236. Boulder: Westview Press.

Sachs, Ignacy. 1977. *Environment and Development—A New Rationale for Domestic Policy Formulation and International Cooperation Strategies.* Ottawa: Canadian International Development Agency and Environment Canada.

———— 1984. *Development and Planning* trans. by Peter Fawcett. Cambridge: Cambridge University Press.

Sachs, Jeffrey. 1995. "Consolidating Capitalism." *Foreign Policy* 98:51–64.

Sachs, Wolfgang. 1992. "Environment." In Wolfgang Sachs, ed., *The Development Dictionary: A Guide to Knowledge as Power*, pp. 26–37. London: Zed Books.

———— ed. 1993. *Global Ecology: A New Arena of Political Conflict.* London: Zed Books.

Sagasti, Francisco R. and Michael E. Colby. 1993. "Eco-Development and Perspectives on Global Change from Developing Countries." In Nazli Choucri, ed., *Global Accord*, pp. 175–203. Cambridge: MIT Press.

Sand, Peter. 1993. "Kaleidoscope: International Environmental Law After Rio." *European Journal of International Law* 4:377–389.

Sands, Philippe, ed. 1994. *Greening International Law.* New York: The New Press.

———— et al., eds. 1994. *Documents in International Environmental Law.* Manchester and New York: Manchester University Press.

Saurin, Julian. 1996. "International Relations, Social Ecology and the Globalisation of Environmental Change." In John Vogler and Mark Imber, eds., *The Environment and International Relations*, pp. 77–98. London: Routledge.

Schachter, Oscar. 1991. "The Emergence of International Environmental Law, *Journal of International Affairs* 44:457–493.

Schmidheiny, Stephan, ed. 1992. *Changing Course*. Cambridge, MA: MIT Press.

Schmidt, Markus G. 1989. *Common Heritage or Common Burden?* Oxford: Clarendon Press.

Schrijver, Nico. 1997. *Sovereignty Over Natural Resources: Balancing Rights and Duties*. Cambridge: Cambridge University Press.

SCOPE. 1971. *Global Environmental Monitoring: A Report Submitted to the United Nations Conference on the Human Environment, Stockholm 1972*. Sweden: SCOPE.

———— 1972. "Environmental Problems in the Developing Countries: Basic Issues." Summary of the report prepared by the Working Party convened by SCOPE in co-operation with the secretariat of the Conference. A/CONF.48/10. Annex II of "Educational, Informational, Social and Cultural Aspects of Environmental Problems: Report by the Secretary-General." A/CONF.48/9. In U.S. Department of State. *Documents for the UN Conference on the Human Environment Stockholm, June 5–16, 1972*. National Technical Information Service, U.S. Department of Commerce, vols. 1–3.

Searle, John R. 1995. *The Construction of Social Reality*. New York: Free Press.

Sebenius, James. 1993. "The Law of the Sea Conference: Lessons for Negotiations to Control Global Warming." In Gunnar Sjöstedt, ed., *International Environmental Negotiation*, pp. 184–216. Newbury Park, CA: Sage.

Shafer, Sarah L. and Alexander B. Murphy. 1998. "The Territorial Strategies of IGOs: Implications for Environment and Development." *Global Governance* 4:257–274.

Shibata, Akiho. 1994. "International Law-Making Process in the United Nations: Comparative Analysis of UNCED and UNCLOS III." *California Western International Law Journal* 24:17–53.

Sikkink, Kathryn. 1993. "Human Rights, Principled Issue Networks, and Sovereignty in Latin America." *International Organization*. 47:411–441.

———— 1995. "Nongovernmental Organizations and Transnational Issue Networks in International Politics." *ASIL Proceedings* (April 8):416–420.

Sjöstedt, Gunnar. 1994. "Issue Clarification and the Role of Consensual Knowledge in the UNCED Process." In Bertram I. Spector et al., eds., *Negotiating International Regimes: Lessons Learned from the United Nations Conference on Environment and Development (UNCED)*, pp. 63–86. London: Graham and Trotman/ Martinus Nijhoff.

———— et al. 1994. "The Dynamics of Regime-building Negotiations." In Bertram I. Spector et al., eds., *Negotiating International Regimes: Lessons Learned from the United Nations Conference on Environment and Development (UNCED)*, pp. 3–19. London: Graham and Trotman/Martinus Nijhoff.

Skodvin, Tora. 2000. "The Intergovernmental Panel on Climate Change." In Steinar Andresen et al., eds., *Science and Politics in International Environmental Regimes: Between Integrity and Involvement*, pp. 146–180. Manchester: Manchester University Press.

Smets, Henri. 1994. "The Polluter Pays Principle in the Early 1990s." In Luigi

Campiglio et al., eds., *The Environment After Rio: International Law and Economics*, pp. 131–147. London: Graham and Trotman/Martinus Nijhoff.

Smith, Jackie, Charles Chatfield and Ron Pagnucco, eds. 1997. *Transnational Social Movements and Global Politics: Solidarity Beyond the State*. Syracuse, NY: Syracuse University Press.

Snidal, Duncan. 1985. "The Limits of Hegemonic Stability Theory." *International Organization* 39:379–614.

Sohn, Louis B. 1973. "The Stockholm Declaration on the Human Environment." *Harvard International Law Journal* 14:423–515.

South Centre. 1991. *Report of the South Centre on Environment and Development: Towards a Common Strategy for the South in the UNCED Negotiations and Beyond*. South Centre, The Follow-up Office of the South Commission.

Spector, Bertram I. et al., eds. 1994. *Negotiating International Regimes: Lessons Learned from the United Nations Conference on Environment and Development (UNCED)*. London: Graham and Trotman/Martinus Nijhoff.

Speth, James Gustave. 1990. "Toward a North South Compact for the Environment." *Environment* 32:16–20, 41–43.

Sprinz, Detlef F. 1999. "Research on the Effectiveness of International Environmental Regimes: A Review of the State of the Art." Paper presented at the 1999 International Studies Association Convention, Washington.

——— and Tapani Vaahtoranta 1994. "The Interest-Based Explanation of International Environmental Policy." *International Organization* 48:77–105.

Spruyt, Hendrik. 1994. *The Sovereign State and Its Competitors*. Princeton, NJ: Princeton University Press.

Stavins, Robert and Thomas Grumbly. 1993. "The Greening of the Market: Making the Polluter Pay." In Will Marshall and Martin Schram, eds., *Mandate for Change*, pp. 197–216. New York: Berkley Books for the Progressive Policy Institute and the Democratic Leadership Council.

Stavins, Robert and Bradley W. Whitehead. 1992. "Dealing With Pollution." *Environment* 34:7–11, 29–42.

Stiglitz, Joseph E. 1998. "More Instruments and Broader Goals: Moving Toward the Post-Washington Consensus." World Institute for Development Economics Research, United Nations University, Annual Lecture #2, Helsinki, January 7.

——— 1999. "Whither Reform? Ten Years of the Transition." Keynote Address of the World Bank Annual Bank Conference on Development Economics, Washington, D.C., April 28–30.

Strong, Maurice. 1975. "Progress or Catastrophe: Whither Our World." *Environmental Conservation* 2:83–88.

——— 1977. "The International Community and the Environment." *Environmental Conservation* 4:165–172.

——— 1992. "Launching an Era of International Environmental Negotiations." An interview in Steve Lerner. *Beyond the Earth Summit: Conversations with Advocates of Sustainable Development*, 21–33. Bolinas, CA: Commonweal.

——— 2000. *Where in Earth Are We Going?* Toronto: Knopf Canada.

Susskind, Lawrence E. 1994. *Environmental Diplomacy: Negotiating More Effective Global Agreements*. New York: Oxford University Press.

Thacher, Peter S. 1973. "ICSU's Role in the Light of the Stockholm Conference on the Human Environment." *ICSU Bulletin* 29:10–19.

Thomas, Caroline. 1992. *The Environment in International Relations*. London: Royal Institute of International Affairs.

———, ed. 1994. *Rio: Unravelling the Consequences*. London: Frank Cass.

Thompson, Andrew R. 1972. "Legal Responses to Pollution Problems—Their Strengths and Weaknesses." *Natural Resources Journal* 12:227–241.

Thompson, Koy. 1993. "The Rio Declaration on Environment and Development." In Michael Grubb et al., eds., *The Earth Summmit Agreements*, pp. 85–95. London: Earthscan.

Timberlake, Lloyd. 1989. "The Role of Scientific Knowledge in Drawing up the Brundtland Report." In Steiner Andresen and Willy Østreng, eds., *International Resource Management: The Role of Science and Politics*, pp. 117–123. London: Belhaven Press.

Tolba, Mostafa K. 1987. *Sustainable Development: Constraints and Opportunities*. London: Butterworth.

Turner, R. Kerry. 1995. "Environmental Economics and Management." In Timothy O'Riordan, *Environmental Science for Environmental Management*, pp. 30–44. Essex: Longman Scientific and Technical.

Underdal, Arild. 2000a. "Science and Politics: The Anatomy of an Uneasy Partnership." In Andresen et al., eds. *Science and Politics in International Environmental Regimes: Between Integrity and Involvement*, pp. 1–21. Manchester: Manchester University Press.

——— 2000b. "Comparative Conclusions." In Andresen et al., eds. *Science and Politics in International Environmental Regimes: Between Integrity and Involvement*, pp. 181–201. Manchester: Manchester University Press.

United Nations. 1971. *Report of the Secretary General: Development and Environment (subject area V). Conference on the Human Environment*. New York: UN, Dec. 22, 1971, A/Conf.48/10.

——— 1972a. *Report of the United Nations Conference on the Human Environment*. A/CONF.48/14/REV.1 (United Nations).

——— 1972b. "Stockholm Declaration of the United Nations Conference on the Human Environment." A/CONF.48/14 and Corr.1. In United Nations 1972a.

——— 1989. "United Nations Conference on Environment and Development." United Nations General Assembly Resolution 44/228, December 22.

——— 1993. *Proceedings of the United Nations Conference on Trade and Development: Eighth Session (Cartagena de Indias, Columbia February 8–25, 1992)*. New York: United Nations.

——— 1994. "Agreement Relating to the Implementation of Part XI of the United Nations Convention on the Law of the Sea of December 10, 1982, with Annex, adopted at New York, July 28, 1994." (UNGA A/RES/48/263). Reproduced in "Message from the President of the United States and Commentary Accompanying the United Nations Convention on the Law of the Sea and the Agreement Relating to the Implementation of the Part XI Upon Their Transmittal to the United States Senate for its Advice and Consent." Reprinted in *The Georgetown International Environmental Law Review* 7:77–190ff.

UN Commission on Sustainable Development (UNCSD). 1997. Report of the Secretary General, "Rio Declaration on Environment and Development: Application and Implementation." UNCSD, Fifth Session (April 7–25). E/CN.17/1997/8 (Feb. 10, 1997).

UNCSD. 1996. Report of the Secretary-General, "Financial Resources and Mechanisms, Addendum." UNCSD, Fourth Session (April 18–May 3). E/CN.17/1996/4/Add.1 (Feb. 22, 1996).

United Nations Environment Programme (UNEP). 1981. *Defense of the Earth: The Basic Texts on Environment: Founex. Stockholm. Cocoyoc.* Nairobi: United Nations Environment Programme.

—— *International Register of Environmental Treaties.*

UN General Assembly (UNGA). 1980. "Development and International Cooperation: Report of the 2nd Committee, Part II." A/35/592/Add.1. November 27:11–12.

—— 1994. "Report of the Secretary General: Consultations of the Secretary-General on outstanding issues relating to the deep seabed mining provisions of the United Nations Convention on the Law of the Sea." UNGA A/48/950 (June 9).

—— 1997. "Programme for the Further Implementation of Agenda 21." UNGA A/RES/S-19/2.

United States Department of State. 1972. *Documents for the UN Conference on the Human Environment Stockholm, June 5–16, 1972* by the National Technical Information Service, U.S. Department of Commerce.

Verhoeve, Barbara, et al. 1992. *Maastricht and the Environment.* London: Institute for European Environmental Policy.

Victor, David G. 1995. "On Writing Good Histories of Climate Change and Testing Social Science Theories." *Climatic Change* 29:363–369.

—— and Julian E. Salt. 1994. "From Rio to Berlin . . . Managing Climate Change." *Environment* 36(10):6–15, 25–32.

—— 1995. "Keeping the Climate Treaty Relevant: An Elaboration." Mimeo (Berlin, April 3).

Ward, Barbara and René Dubos. 1972. *Only One Earth: The Care and Maintenance of a Small Planet.* An Unofficial Report Commissioned by the Secretary-General of the United Nations Conference on the Human Environment, Prepared with the Assistance of a 152-Member Committee of Corresponding Consultants in 58 Countries. New York: Norton.

Weale, Albert. 1992. *The New Politics of Pollution.* Manchester: Manchester University Press.

Weber, Steve. 1994. "Origins of the European Bank for Reconstruction and Development." *International Organization* 48:1–38.

Wendt, Alexander. 1987. "The Agent-Structure Problem in International Relations Theory." *International Organization* 41:335–370.

—— 1992. "Anarchy Is What States Make of It: The Social Construction of Power Politics." *International Organization* 46:391–425.

—— 1994. "Constructing International Politics." *International Security* 20:71–81.

—— 1998. "On Constitution and Causation in International Relations." In Tim Dunne, Michael Cox and Ken Booth, eds., *The Eighty Years' Crisis: International Relations 1919–1999,* pp. 101–117. Cambridge: Cambridge University Press.

———— and Raymond Duvall. 1989. "Institutions and International Order." In Ernst-Otto Czempiel and James N. Rosenau, eds., *Global Changes and Theoretical Challenges*, pp. 51–73. Lexington: Lexington Books.

Wettestad, Jorgen. 1999. *Designing Regime Effectiveness: The Key Conditions.* Cheltenham, UK: Edward Elgar.

———— 2000. "The ECE Convention on Long-Range Transboundary Air Pollution: From Common Cuts to Critical Loads." In Andresen et al., eds., *Science and Politics in International Environmental Regimes: Between Integrity and Involvement*, pp. 95–121. Manchester: Manchester University Press.

Williams, Marc. 1993. "Re-articulating the Third World Coalition: The Role of the Environmental Agenda." *Third World Quarterly* 14:7–29.

———— 1996. "International Political Economy and Global Environmental Change." In John Vogler and Mark F. Imber, eds., *The Environment and International Relations*, pp. 41–58. London: Routledge.

Williamson, John. 1990. "The Progress of Policy Reform in Latin America." *Policy Analyses in International Economics*, no. 28. Washington, D.C.: Institute for International Economics.

————1993. "Democracy and the 'Washington Consensus.' " *World Development* 21:1329–1336.

Wirth. Timothy E. 1996. "Remarks by the Honorable Timothy E. Wirth, Under Secretary for Global Affairs, on behalf of the United States of America before the Second Conference of the Parties Framework Convention on Climate Change." Geneva, Switzerland. July 17.

WMO/UNEP. 1992. *United Nations Framework Convention on Climate Change.* Geneva: WMO/UNEP Information Unit on Climate Change.

Woodrow Wilson Center. 1996. *Environmental Change and Security Project.* Washington, D.C.: Woodrow Wilson Center, issue 2.

Woods, Ngaire. 1995. "Economic Ideas and International Relations: Beyond Rational Neglect." *International Studies Quarterly* 39:161–180.

The World Bank. 1990. *The World Bank and the Environment.* Washington, D.C.: World Bank.

———— 1992a. *The World Bank and The Environment.* Washington, D.C.: World Bank.

———— 1992b. *World Bank Development Report, 1992: Development and Environment.* New York: Oxford University Press.

———— 1994. *Making Development Sustainable: The World Bank Group and the Environment.* Washington, D.C.: World Bank.

World Bank, UNDP, and UNEP. 1992. *Global Environment Facility: The Pilot Phase and Beyond.* Working Paper Series, Number 1. Washington, D.C.: GEF.

The World Commission on Environment and Development (WCED). 1987. *Our Common Future.* Oxford: Oxford University Press.

World Trade Organization (WTO). 2000. "Committee Tackles Market Access Issues." *Trade and Environment Bulletin* 32 (March 17).

Worster, Donald. 1993. "The Shaky Ground of Sustainability." In Wolfgang Sachs, ed., *Global Ecology: A New Arena of Political Conflict*, pp. 133–145. London: Zed Books.

Wynter, Marie. 1999. "The Use of Market Mechanisms in the Shrimp-Turtle Dispute: The WTO's Response." In Klaus Bosselmann and Benjamin J. Richardson, eds., *Environmental Justice and Market Mechanisms: Key Challenges for Environmental Law and Policy*, pp. 169–188. The Hague: Kluwar Law International.

Wysham, Daphne. 1994. "Ten-to-One Against: Costing People's Lives for Climate Change." *The Ecologist* 24:204–206.

Yearbook of the United Nations (YUN). Various years.

Yearley, Steven. 1992. "Green Ambivalence About Science: Legal-Rational Authority and the Scientific Legitimation of a Social Movement." *British Journal of Sociology* 43:511–532.

Yee, Albert S. 1996. "The Causal Effects of Ideas on Policies." *International Organization* 50:69–108.

Young, Oran. 1989. "The Politics of International Regime Formation: Managing Natural Resources and the Environment." *International Organization* 43:349–375.

——— 1994. *International Governance: Protecting the Environment in a Stateless Society*. Ithaca: Cornell University Press.

——— 1995. *Global Governance: Drawing Insights from Environmental Experience*. Hanover, NH: Occasional Paper from the Dickey Center, Dartmouth College.

——— 1996. "Institutional Linkages in International Society: Polar Perspectives." *Global Governance* 2:1–24.

——— (ed.) 1999. *The Effectiveness of International Environmental Regimes: Causal Connections and Behavioral Mechanisms*. Cambridge, MA: MIT Press.

——— and Gail Osherenko, eds. 1993. *Polar Politics: Creating International Environmental Regimes*. Cornell: Cornell University Press.

Zucker, Lynne G. 1991. "The Role of Institutionalization in Cultural Persistence." In Walter W. Powell and Paul J. DiMaggio, eds., *The New Institutionalism in Organizational Analysis*, pp. 83–107. Chicago: University of Chicago Press.

Zürn, Michael. 1998. "The Rise of International Environmental Politics: A Review of Current Research." *World Politics* 50:617–649.

FORMAL INTERVIEWS REFERENCED

James Bruce, June 18, 1996. (Ottawa)

William Clark, February 20 and February 25, 1997. (Telephone to Center for Science and International Affairs, Harvard, Cambridge, MA)

Jim MacNeill, December 11, 1995. (Toronto)

Gordon McBean, June 28, 1996. (Toronto)

Alden Meyer, April 6, 1995. (Berlin)

R.E. Munn, September 28, 1995. (Toronto)

Robert Stavins, August 2, 1996. (Center for Science and International Affairs, Harvard University, Cambridge, MA)

Maurice Strong, February 19, 1996. (Toronto)

Jeffrey Watson, June 17, 1996. (Ottawa)

Timothy Wirth, March 13, 1997. (Telephone to Washington, D.C.)